Ecological Studies

Analysis and Synthesis

Volume 254

Ecological Studies is Springer's premier book series treating all aspects of ecology. These volumes, either authored or edited collections, appear several times each year. They are intended to analyze and synthesize our understanding of natural and managed ecosystems and their constituent organisms and resources at different scales from the biosphere to communities, populations, individual organisms and molecular interactions. Many volumes constitute case studies illustrating and synthesizing ecological principles for an intended audience of scientists, students, environmental managers and policy experts. Recent volumes address biodiversity, global change, landscape ecology, air pollution, ecosystem analysis, microbial ecology, ecophysiology and molecular ecology.

William D. Newmark

The Kihansi Spray Toad *(Nectophrynoides asperginis)*: Crisis Biology in the 21st Century

Springer

William D. Newmark
Natural History Museum of Utah
University of Utah
Salt Lake City, UT, USA

ISSN 0070-8356 ISSN 2196-971X (electronic)
Ecological Studies
ISBN 978-3-031-89598-2 ISBN 978-3-031-89599-9 (eBook)
https://doi.org/10.1007/978-3-031-89599-9

This Springer imprint is published by the registered company Springer Nature Switzerland AG
The registered company address is: Gewerbestrasse 11, 6330 Cham, Switzerland

The Kihansi Spray Toad (*Nectophrynoides asperginis*). (Photograph by Timothy Herman)

Acknowledgments

I would like to thank the many individuals who so generously contributed their time, input, and assistance to the conservation of the Kihansi Spray Toad including Paola Agostini, George Amato, Dinesh Aral, Anne Baker, Rutabanzibwa Bamwenda, Bonaventure Baya, Val Beasley, John Behler, Dominico Benedicto, Alyssa Borek, Don Boyer, Jim Breheny, Matthew Buff, Kurt Buhlmann, Yesmeana Butler, Faustina Chande, Allan Channing, John Chastian, Ladisy Chengula, Don Church, James Collins, Diane Davidson, Peter Dewees, Daniel Doak, Nilda Ferrer, Siobain Finlow-Bates, Yolan Friedman, Claude Gascon, John Gerstle, James Gibbs, Ezekiel Goboro, Ana Gonzalez, Louis Gross, Carmen Guzman, Svein Haarklau, Wyn Hall, Chris Hanley, Peter Hawkes, Timothy Herman, Indu Hewawasam, Rafik Hirji, Bill Holmstrom, Kim Howell, Radhia Ideva, Dennis Ikanda, Nevena Ilieva, Anders Jensen, Asha Johnson, L Kaaya, P Kalenga, Gayatri Kanungo, Bakari Kaoneka, William Karesh, Herman Keraryo, Julius Keyya, Fadhila Khatibu, Deusdedit Kibassa, Jane Kibbassa, Amina Kibola, Idris Kikula, Dominico Kilemo, JF Killenga, E Kimaro, Blake Klocke, Jessi Krebs, PM Kusolwa, George Ledec, Sam Lee, James Lewis, Jan Lindemark, Stephen Linter, Karen Lips, Magda Lovei, Thomas Lovejoy, Jon Lovett, H Lyaruu, Anna Maembe, Flora Magige, Mariam Makange, Joshua Malago, Hamdun Mansur, Gerry Maranteli, Nicodems Marcus, Charles Masunzu, Hamdun Masur, Faith-Lucy Matumbo, Dee McAloose, John McIntire, Simon Mduma, Jean-Roger Mercier, David Miller, Gerald Misinzo, Eblate Mjingo, Nassoro Mohamed, Raphael Mollel, Robin Moore, Kerry Morrison, Charles Msuya, Samuel Mtoka, Devolent Mtui, David Muginya, Ruzika Muheto, Severinus Mutagwaba, Henrick Mwalongo, A Mwambene, Dismass Mwikila, Cuthbert Nahonyo, Aaron Nanyaro, Henry Ndangalasi, Montague Neate-Clegg, Wilirk Ngalason, Magnus Ngoile, Maura O'Connor, Andrew Odum, Graeme Patterson, Allan Pessier, Parson Peter, Valentine Peter, Jennifer Pramuk, CH Quinn, Chelsae Radell, L Ramos, Raphael Mwalyosi, Brooke Reeve, Corrine Richards-Zawaki, Alfan Rija, Francis Sabuni, Wilfred Sarunday, Robert Schlotterer, Jason Searle, Wynona Shellabarger, Gloria Sindano, Pritpal Soorae, Justin Sorenson, Thomas Stanley, Simon Stuart, TB Suya, RA Tamatamah, Jonathan Tangwa, James Taplin,

Jeremy Thompson, Valorie Titus, Elizabeth Townsend, Tracey Tuberville, Peter Ugomba, Felista Urasa, Patrick Valimba, Tom Walton, Ché Weldon, John Wood, Innocent Zilihona, and Kevin Zippel.

Competing Interests The author has no competing interests to declare that are relevant to the content of this manuscript.

Contents

Abbreviations

CBSG	Conservation Breeding Specialist Group
DOE	Department of Energy
GEF	Global Environment Facility
IREM	Immediate Rescue Emergency Measures Project
IUCN	International Union for Conservation of Nature
KCCMP	Kihansi Catchment Conservation and Management Project
KfW	KfW Development Bank
LKEMP	Lower Kihansi Environmental Management Project
LKHP	Lower Kihansi Hydropower Project
MEM	Ministry of Energy and Minerals
MNRT	Ministry of Natural Resources and Tourism
NEMC	National Environmental Management Council
TNO	Netherlands Organization for Applied Scientific Research
NORAD	Norwegian Agency for Development Cooperation
SIDA	Swedish International Development Cooperation Agency
SABS	South African Bureau of Standards
SSC	Species Survival Committee
SUA	Sokoine University of Agriculture
SUNY	State University of New York
TANESCO	Tanzania Electric Supply Company
TAWIRI	Tanzania Wildlife Research Institute
TEDEP	Tanzania Energy Development and Access Expansion Project
TPRI	Tropical Pest and Research Institute
UDSM	University of Dar es Salaam
WD	Wildlife Division

Captive Breeding and Reintroduction of Globally Threatened Species

1

Abstract

The captive breeding and the reintroduction of threatened species are among the most challenging conservation interventions. To date, 23 globally threatened, endangered, and at-risk anuran species have been captive-bred and reintroduced. Of these 23 anuran species, 5 are tropical. Based on meta-analyses conducted since the late 1980s, translocation success of amphibians and herpetofauna is 14% lower on average than translocation success for mammals, birds, and combined taxa. This is very likely a result of the emergence of chytridiomycosis and its interaction with other risk factors including habitat loss, climate change, toxic compounds, and invasive species. Previous meta-analyses of terrestrial vertebrate taxa have also revealed that translocation success for many taxa is higher among (1) large than small populations; (2) wild-sourced than captive-bred populations; (3) populations translocated to the core of their historic range; (4) populations translocated to sites where habitat is good to excellent; and (5) soft than hard releases. This book describes the species recovery program that was designed and implemented to conserve the Kihansi Spray Toad *Nectophrynoides asperginis* an Afrotropical and extinct-in-the-wild species.

The captive breeding and reintroduction of globally threatened species are among the most challenging conservation interventions and often represent the final tool in a comparatively small tool box to conserve rapidly declining species in the wild. Such interventions are not only technically challenging but are also expensive in terms of human and financial resources. Moreover, for many threatened species, little is often known about their biology and husbandry. Yet in spite of these challenges, captive breeding combined with reintroduction is becoming increasingly

W. D. Newmark, *The Kihansi Spray Toad* (Nectophrynoides asperginis): *Crisis Biology in the 21st Century*, Ecological Studies 254, https://doi.org/10.1007/978-3-031-89599-9_1

necessary to save many globally threatened species due to the ever-increasing impact of human activities on biodiversity worldwide (Crates et al. 2023; Smith et al. 2023).

There is increasing evidence that planet Earth is now entering the Sixth Mass Extinction event (Ceballos et al. 2015, 2017; IPBES 2019; Cowie et al. 2022). Current extinction rates of species are estimated to be 100 to 1000 times higher than average background rates (Pimm et al. 2014; Ceballos et al. 2015; Cowie et al. 2022). Yet, the current global loss of species differs from earlier mass extinction events in that species extinctions are a result of human activities rather than natural phenomenon (IPBES 2019; Cowie et al. 2022).

Among vertebrate species, one-fifth of all species are classified by IUCN as threatened (Hoffmann et al. 2010). And among vertebrates, amphibians are the most threatened class, of which 41% of species are classified as threatened followed by cartilaginous fish (33%), mammals (25%), reptiles (22%), bony fish (15%), and birds (13%) (Hoffmann et al. 2010). Between 1980 and 2004, amphibians also declined the most of any vertebrate class in terms of the change over time in their IUCN Red List conservation status, with a decline of 0.14% per year in their Red List category index (Hoffmann et al. 2010).

The dramatic global decline in amphibian species was first highlighted in 1989 at the First World Congress on herpetology and then subsequently quantified over the following decade and a half (Houlahan et al. 2000; Stuart et al. 2004). In these earlier global assessments, the decline and extinction of amphibian species worldwide were attributed largely to habitat loss, overexploitation, and other "enigmatic" factors (Stuart et al. 2004). In the early 2000s, approximately one-half (48%) of all declines and extinctions of amphibian species were categorized as "enigmatic" (Stuart et al. 2004).

Over the last two decades with the identification of the disease chytridiomycosis (Berger et al. 1998) and the fungi that were responsible *Batrachochytrium dendrobatidis* (Bd) (Longcore et al. 1999) and *Batrachochytrium salamandrivorans* (Bsal) (Martel et al. 2013), many of these previously categorized "enigmatic" declines are now attributed to the chytridiomycosis panzootic (Fisher and Garner 2020). As of 2019, chytrid fungi has been found to infect 54% of all tested species ($n = 1854$) and has been detected at 39% of all tested sites ($n = 9503$) (Fisher and Garner 2020). In a recent meta-analysis, declines in 501 amphibian species worldwide were attributed to chytridiomycosis of which 90 species (18%) are confirmed or presumed extinct and 124 species (25%) have experienced a >90% decline in abundance (Scheele et al. 2019). The amphibian chytridiomycosis panzootic has been described as "the greatest recorded loss of biodiversity attributable to a disease" (Scheele et al. 2019).

As highlighted by the precipitous decline of many amphibian species in undisturbed environments, chytridiomycosis is often the principal driver of the decline of many amphibian species (Skerratt et al. 2007). Yet in many other species, chytridiomycosis can interact with and exacerbate the adverse impacts of other major drivers of amphibian declines including habitat loss, climatic change, and invasive species (Wake and Vredenburg 2008).

1.1 Captive Breeding of Threatened Species

In response to the global decline of amphibians and the emergence of the chytridiomycosis panzootic, there has been an increasing effort over the last two decades to establish captive assurance populations for many threatened amphibian species (Mendelson et al. 2006; Smith et al. 2023). The importance of captive breeding combined with reintroductions as part of a broader strategy to conserve threatened amphibians was highlighted in 2007 in the Amphibian Conservation Action Plan (Gascon et al. 2007). Biologists working in Australia and Central America have pioneered much of this effort (Lewis et al. 2019; Scheele et al. 2021). Yet as Gascon et al. have noted, there is still a considerable mismatch between the scale of the problem and capacity to house threatened amphibian species (Gascon et al. 2012).

Amphibians in comparison to many other vertebrate species possess life history traits that are well-suited for captive breeding. These include short generation time, high fecundity, and small body size. And due to this latter trait, it is easier to house large populations of amphibians in captive breeding facilities than many other vertebrate species (Griffiths and Pavajeau 2008; Harding et al. 2016). Additionally, much of amphibian behavior is hard-wired, and thus pre- and post-release training is not required unlike in many higher vertebrates that possess complex social systems (Griffiths and Pavajeau 2008; Harding et al. 2016).

1.2 Translocation of Threatened Species

Reintroduction biology as a formal discipline is a young yet rapidly growing field (Kleiman 1989; Gipps 1991; Serena 1995; Seddon et al. 2007; Armstrong and Seddon 2008; Ewen et al. 2012; Jachowski et al. 2016). Yet translocation of species for conservation purposes dates back to the mid-1890s to the early 1900s. Between 1895 and 1907, conservationists in New Zealand translocated 474 Kakapo, Little spotted Kiwi, and Brown Kiwi that were sourced from the wild to 11 small islands in the Dusky Sound in New Zealand (Miskelly and Powlesland 2013). Although the reintroductions of the Kakapo and Little spotted Kiwi were initially unsuccessful, the current populations of the Brown Kiwi on three inlands in the Dusky Sound are thought to have descended from the original reintroduction (Miskelly and Powlesland 2013).

In North America, possibly the earliest record of captive breeding and reintroduction of a species for conservation purposes is the American bison. In the early 1900s, a captive breeding program was initiated at the Bronx zoo for the American bison under the aegis of the American Bison Society (Aune and Plumb 2019). This was followed in 1907 by the reintroduction of 15 animals from this captive-bred population to the Wichita Mountains National Forest and Game Reserve in Oklahoma (Kleiman 1989; Aune and Plumb 2019). This reintroduced population combined with subsequent translocations was successful in establishing a self-sustaining population of the American Bison in the Wichita Mountains.

Over the last half century, the number of globally threatened species that have been captive-bred and reintroduced has expanded rapidly. Among mammals, examples of captive-bred and reintroduced globally threatened species include the golden lion tamarin (Kleiman 1989), the Arabian oryx (Stanley Price 1989), Père David's deer (Jiang et al. 2000), the red wolf (Seal 1991), the eastern barred bandicoot (Backhouse et al. 1995), the black-footed ferret (Dobson and Lyles 2000), and the dibbler (Moro 2003). Among birds, prominent examples of captive bred-reintroduced globally threatened species include the Mauritius kestrel (Cade and Jones 1993), the Mauritius fody (Jones et al. 1995), the orange-bellied parrot (Brown et al. 1995), the California condor (Snyder and Snyder 2000), the takahe (Lee and Jamieson 2001), and the Guam rail (Trask et al. 2020).

1.3 Comparison of Translocation Success Among Taxa

To date, the overall success of translocation of species for conservation purposes remains low (Gross et al. 2024). Over the last four decades, more than 20 meta-analyses have been conducted that have compared translocation success of captive-bred and wild-sourced populations. These meta-analyses, conducted both at a global and regional scale, have focused on multiple taxa. Eighteen of these meta-analyses conducted since the late 1980s are summarized in Table 1.1. In this summary, I have highlighted where possible translocation success of threatened, endangered, or at-risk species or where translocations were conducted for conservation purposes.

However, comparing translocation success of populations within and between meta-analyses is complicated by how translocation success is defined. The most frequent definition of translocation success across the 18 meta-analyses was "the establishment of a self-sustaining population." Yet as Germano and Bishop (2009) have noted, this definition of success is dependent on monitoring the released population over a sufficiently long period so as to be able to detect not only survival and growth of released individuals but also reproduction and population growth in the released population. Length of the post-release monitoring does vary considerably between studies incorporated in the meta-analyses presented in Table 1.1. The second most frequent definition of translocation success included in Table 1.1 is based on the original study authors' interpretation of translocation success.

A further factor that complicates comparison of translocation success within and between the meta-analyses presented is variation in terminology surrounding translocation. In 2013 the IUCN Specialist Group for Reintroduction and Invasive developed a series of guidelines to assist conservation practitioners in planning and implementing species translocations. As part of these guidelines, the IUCN Specialist Group for Reintroduction defined key terms associated with species translocation. Translocation is defined as "the intentional movement of living organisms from one area to another," and reintroduction, a subcategory of translocation, as "the intentional movement and release of an organism inside its indigenous range from which it has disappeared" (IUCN/SSC 2013).

Table 1.1 Translocation success of five focal taxa

Focal taxa	Sub-category	Geographic focus	Number of translocations	Success (%)	Definition of success	Survey period	Population of successful vs unsuccessful translocations[a]	Translocation success (%) of wild-sourced vs captive- bred[b]	Source
Combined taxa birds (68%), mammals (32%)	Threatened, endangered, and sensitive species	Australia, Canada, Hawaii, New Zealand, USA	80	44	Self-sustaining population	1973–1986	$\bar{x}$ =160 vs 54	75 vs 38	Griffith et al. (1989)
Combined taxa birds (58%), mammals (42%)	Threatened, endangered, and sensitive species	Australia, Canada, Hawaii, New Zealand, USA	89	53	Self-sustaining population	1973–1993	Birds: $\bar{x}$ = 763 vs 82	Birds: 71 vs 40	Wolf et al. (1996)
Combined taxa mammals (49%), birds (44%), amphibians, reptiles, and invertebrates (7%)	75% of reintroductions were for conservation purposes	Global	116	26	Followed study authors' determination	1979–1998	>100 vs < 100	31 vs 13	Fischer and Lindenmayer (2000)
Combined taxa mammals (67%), birds (14%), amphibians (7%), reptiles (5%), invertebrates (7%)	Threatened species	Western Australia	191	35	Followed study authors' determination	1971–2014	>71 vs < 71	27 vs 59	Morris et al. (2015)

(continued)

Table 1.1 (continued)

Focal taxa	Sub-category	Geographic focus	Number of translocations	Success (%)	Definition of success	Survey period	Population of successful vs unsuccessful translocations[a]	Translocation success (%) of wild-sourced vs captive- bred[b]	Source
Combined taxa mammals (22%), birds (22%), herpetofauna (21%), fish (21%), invertebrates (14%)	Threatened species comprised 25% of taxa	Global	554	54	Followed study authors' determination	2005–2016			Bubac et al. (2019)
Combined taxa mammals (75%), birds (18%), amphibians (4%), reptiles (3%)	Threatened species	Australia	116	47	Absence of failure or persisted >3 or >5 years	1880s–2000s		48 vs 35	Short (2009)
Mammals	Threatened species	Australia	82	56	Absence of failure or persisted >3 or >5 years	1880s–2000s			Short (2009)
Mammals	Threatened species	Australia	10	50	Extant after 8 years	1998–2008	>65 vs <65		Moseby et al. (2011)
Mammals	Carnivores of which 54% were threatened species	Global	536	39	Followed study authors' determination	1890s–2020s			Stepkovitch et al. (2022)

Birds	Threatened species	Australia	25	32	Followed study authors' determination	1880s–2000s	>50 vs <50	43 vs 42	Short (2009)
Birds	Threatened and at-risk taxa comprise 78% of translocated taxa	New Zealand	495	44	Population established and persists	1863–2012		43 vs 43	Miskelly and Powlesland (2013)
Herpetofauna reptiles (80%), amphibians (20%)	Threatened species	Global	25	19	Population established and persists	1962–1990			Dodd and Seigel (1991)
Herpetofauna reptiles (55%), amphibians (45%)	Many species were threatened, but their numbers were not reported	Global	85	42	Long-term population persistence	1991–2006	Amphibians: >1000 vs <1000		Germano and Bishop (2009)
Herpetofauna reptiles (89%), amphibians (10%)	84% were threatened or at-risk taxa	New Zealand	74	8; 47 (dependent on definition of success)	Population established and persists; dependent on time since release and life-history traits	1984–2008			Miller et al. (2014)
Amphibians	Threatened species	Australia	9	11	Absence of failure or persisted >3 or >5 years	1880s–2000s			Short (2009)

(continued)

Table 1.1 (continued)

Focal taxa	Sub-category	Geographic focus	Number of translocations	Success (%)	Definition of success	Survey period	Population of successful vs unsuccessful translocations[a]	Translocation success (%) of wild-sourced vs captive- bred[b]	Source
Amphibians	24% were threatened species	Global	21	62	Self-sustaining population	1970–2006			Griffiths and Pavajeau (2008)
Amphibians	43% were threatened species	Global	37	38	Self-sustaining population	1970–2014			Harding et al. (2016)
Amphibians	Threatened species	Australia	15	20	Evidence of reproduction for several years post-release	1998–2021			Scheele et al. (2021)

Data are from 18 meta-analyses published between 1989 and 2022. This summary is focused largely on translocation success of internationally or nationally endangered, threatened, and at-risk species or taxa translocated for conservation purposes

[a]Cutoff value between “large” and “small” founder populations

[b]Includes both threatened and non-threatened species

Notwithstanding these differences in definitions and terminology, several important patterns emerge from this compilation of previous meta-analyses. Firstly, translocation success has consistently been found across all taxa to be higher among large than among small founder populations. This pattern is consistent with earlier reviews of translocation success (Seddon et al. 2014). The minimum reported threshold value separating successful from unsuccessful translocations was largest for amphibians (>1000), followed by combined taxa (>71), mammals (>65), and birds (>50) (Table 1.1).

Secondly, translocation success for mammals and to a lesser degree birds was higher for wild-sourced populations than for captive-bred populations but varied little among other taxa (Table 1.1). This pattern again is consistent with earlier reviews (Jule et al. 2008; Seddon et al. 2014). For mammals and combined taxa (predominantly mammals and birds), translocation success was consistently higher in wild-sourced than in captive-bred populations across most meta-analyses. Among birds, translocation success was higher in wild-sourced than captive-bred populations in one of three meta-analyses and did not differ in the other two meta-analyses (Table 1.1). Among herpetofauna and amphibians, only one of the seven meta-analyses compared translocation success between wild-sourced and captive-bred populations. In this meta-analysis no difference in translocation success was detected between wild-sourced and captive-bred amphibian populations (Table 1.1).

Thirdly, translocation success has been found in one or more meta-analyses to be higher for mammals and birds when translocated to the core of their historic range, to sites where habitat quality is good to excellent (Griffith et al. 1989; Wolf et al. 1996), and among mammals, birds, and reptiles higher in soft in comparison to hard releases (Resende et al. 2021). However due to the few meta-analyses that have examined the effects of habitat quality, location of the release site, and soft in comparison to hard releases on translocation success, these data are not summarized in Table 1.1.

Fourthly, average translocation success across terrestrial classes of vertebrates over the last three decades has been low with an overall average success rate of 37% (range 8–62%) across the five focal classes. This result is also consistent with earlier reviews of translocation success (Armstrong and Seddon 2008; Seddon et al. 2014). Yet, translocation success does vary among classes. Highest average translocation success has been reported for mammals (overall mean = 45%, range 39–56%), followed by combined taxa (overall mean = 43%, range 26–54%), birds (overall mean = 38%, range 32–44%), amphibians (overall mean = 33%, range 11–62%), and herpetofauna (overall mean = 23%, range 8–42%) (Fig. 1.1).

Notably the overall mean translocation success for amphibians and herpetofauna combined is 14% lower than the overall mean translocation success for mammals, birds, and combined taxa (Fig. 1.1). This lower mean translocation success for amphibians and herpetofauna relative to mammals, birds, and combined taxa is very likely due to the multiple stressors and their additive effects that particularly amphibians face in the wild including chytridiomycosis, habitat loss and fragmentation, climate change, pesticides and contaminants, and exotic species (Wake and Vredenburg 2008; Hof et al. 2011; Wake 2012; Campbell Grant et al. 2020).

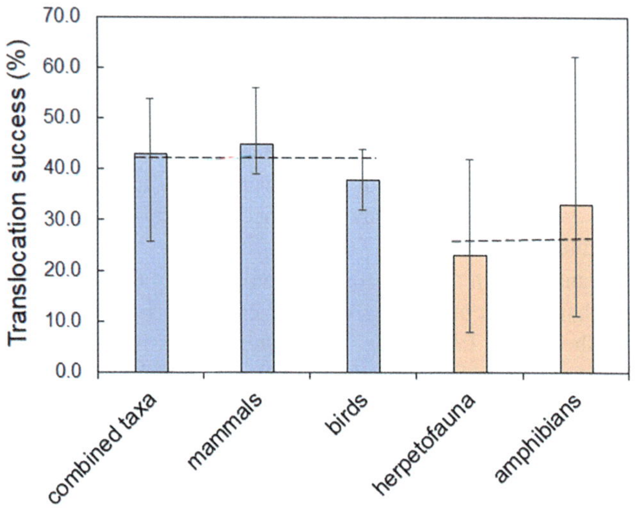

Fig. 1.1 Mean (± range) translocation success by focal taxa across 18 meta-analyses published between 1989 and 2022. Horizontal dashed lines represent the overall mean for combined taxa, mammals, and birds and for herpetofauna and amphibians. A summary of the meta-analyses is presented in Table 1.1

1.4 Captive Breeding: Reintroduction of Anurans

Over the last three decades, 43 anuran species have been captive-bred and reintroduced including the Kihansi Spray Toad (Harding et al. 2016; Hudson 2016; Scheele et al. 2021). Of these 43 species, 2 (5%) are classified by IUCN as extinct in the wild, 7 (16%) as critically endangered, 5 (21%) as endangered, 8 (19%) as vulnerable, 1 (2%) as near threatened, and 20 (47%) as least concerned. Thus, approximately one-half of all captive-bred and reintroduced anuran species over the last three decades have been non-threatened species.

Of the 23 globally threatened, endangered, or at-risk anuran species that have been captive-bred and reintroduced to the wild (Harding et al. 2016; Hudson 2016; Scheele et al. 2021), five species are tropical (Table 1.2), highlighting the under-representation of tropical anuran species in species recovery programs over the last three decades. This pattern further highlights previous observations that although anuran species diversity is highest in the tropics (Jenkins et al. 2013) and is at greatest risk of extinction, numbers of reintroductions of captive-bred anuran species greatly lag behind that of temperate anurans (Tapley et al. 2015; Harding et al. 2016).

1.5 Reintroduction of Tropical Anuran Species

To date the success of reintroductions of tropical anuran species is lower than that among their temperate anurans (Fig. 1.2) although this may be in part a result of the fewer numbers of reintroductions of tropical than temperate anuran species.

Table 1.2 Captive-bred and reintroduced globally threatened, endangered, and at-risk anuran species and their geographic range

Species	Geographic range
Mallorcan midwife toad *Alytes muletensis*	Temperate
Wyoming toad *Bufo baxteri*	Temperate
Houston toad *Bufo houstonensis*	Temperate
Romer's treefrog *Chirixalus romeri*	Temperate
White-bellied frog *Geocrinia alba*	Temperate
Orange-bellied frog *Geocrinia vitellina*	Temperate
Large crested toad *Incilius cristatus*	Tropical
Hamilton's frog *Leiopelma hamiltoni*	Temperate
Maud Island frog *Leiopelma pakeka*	Temperate
Mountain chicken frog *Leptodactylus fallax*	Tropical
Gopher frog *Lithobates capito*	Temperate
Chiricahua leopard frog *Lithobates chiricahuensis*	Temperate
Golden bell frog *Litoria aurea*	Temperate
Booroolong frog *Litoria booroolongensis*	Temperate
Armored frog *Litoria lorica*	Tropical
Spencer's river tree frog *Litoria spenceri*	Temperate
Verreaux's tree frog *Litoria verreauxii*	Temperate
Kihansi spray toad *Nectophrynoides asperginis*	Tropical
Puerto Rican crested toad *Peltophryne lemur*	Tropical
Northern Corroboree frog *Pseudophryne pengilleyi*	Temperate
Corroboree frog *Pseudophryne corroboree*	Temperate
Iberian frog *Rana iberica*	Temperate
Italian agile frog *Rana latastei*	Temperate

Source: Harding et al. (2016), Hudson (2016), and Scheele et al. (2021)

Of the five tropical anuran species that have been captive-bred and reintroduced, one species (20%), the Kihansi Spray Toad, is classified by IUCN as extinct-in-the wild, two (40%) species as critically endangered, and two (40%) species as endangered. Captive bred-reintroduced tropical anuran species are also more threatened globally on average than their temperate counterparts (Fig. 1.3). This pattern highlights the important and emerging role that captive breeding combined with reintroduction is now required to play to conserve the increasing number of threatened anuran species in the tropics. As Harding et al. (2016) note, "the resurgence of zoos-as-arks is reemerging as essential for the conservation of many anuran species, particularly in tropics as a result of the emergence of infectious diseases."

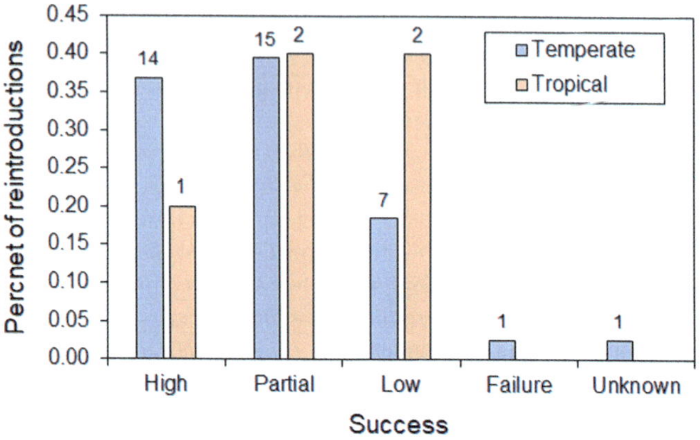

Fig. 1.2 Comparison of the success of reintroduction of captive-bred temperate and tropical anuran species. The labels above bars are number of captive bred-reintroduced species. High success is defined as evidence of reproduction over several generations; partial success as reproduction over one season; low success as evidence of released animals surviving; and failure no evidence of survival. (Source: Harding et al. 2016; Hudson 2016; Scheele et al. 2021)

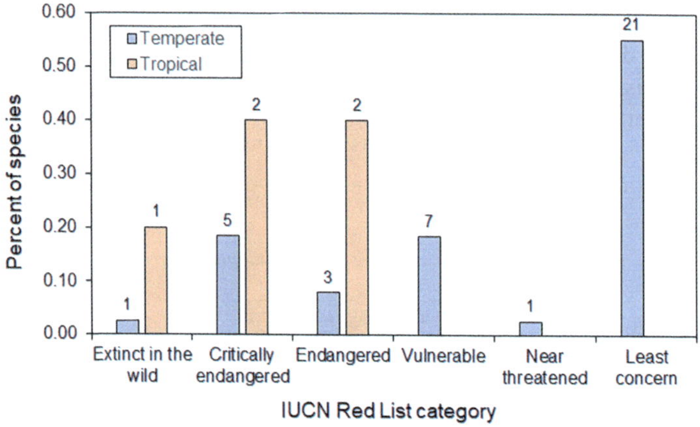

Fig. 1.3 Comparison of the percent of total number of temperate and tropical captive bred – reintroduced anuran species by IUCN Red List category. The labels above bars are numbers of captive bred – reintroduced species. (Source: Harding et al. 2016; Hudson 2016; Scheele et al. 2021)

1.6 Species Recovery Program for the Kihansi Spray Toad

This book describes the species recovery program that was developed and implemented for the Kihansi Spray Toad, an extinct-in-the-wild species. The Kihansi Spray Toad was initially endangered by the construction of a hydroelectric dam located in a remote valley in the Udzungwa Mountains in south central Tanzania in East Africa and then subsequently became extinct-in-the-wild through the putative introduction of chytrid fungus (*Batrachochytrium dendrobatidis*) into the Kihansi Gorge.

In response to national and international concern about the impact of the hydroelectric dam on the Kihansi Spray Toad and its habitat, a species recovery program was developed for the Kihansi Spray Toad. The species recovery program for the Kihansi Spray Toad became a test case for the World Bank and other bilateral donors that co-financed the construction of the Lower Kihansi hydropower facility that their development assistance should not contribute to the extinction of species.

The species recovery program that was developed for the Kihansi Spray was precedent setting in a number of ways. Firstly, the species recovery program for the Kihansi Spray Toad represented the first attempt to captive breed and reintroduce an extinct-in-the-wild tropical amphibian species. Secondly it represented the first captive breeding—reintroduction of an amphibian species in Africa. And thirdly, it represented the first attempt to reintroduce an extinct-in-the-wild tropical amphibian species whose extinction was attributed to chytrid fungus.

The species recovery program that was designed and implemented between 2000 and 2024 following the discovery of the Kihansi Spray Toad in 1996 had, however, to address more than conserving the Kihansi Spray Toad and its critical habitat. The species recovery program had also to design a policy framework in Tanzania for allocation of water for the environment. In the late 1990s, there were no guidelines for allocating water between the environment and human consumption, agriculture, electrical generation, and navigation. Additionally, to ensure the long-term sustainability of the species recovery program, an extensive training program was developed during the first decade to enhance capacity within Tanzania to design and manage multi-sector species recovery programs for threatened species.

Given the multiple threats facing the Kihansi Spray Toad in the late 1990s and early 2000s which included habitat loss, a dramatic reduction in water quantity, a decrease in water quality, changing land use in the watershed, rolling electrical blackouts, and the introduction of chytrid fungus into the Kihansi Gorge, developing a comprehensive species recovery program for the Kihansi Spray Toad that would address these multiple ecological and social challenges was daunting. This book described the more than two decades effort by multiple institutions and individuals, both national and international, to conserve the Kihansi Spray Toad.

References

Armstrong DP, Seddon PJ (2008) Directions in reintroduction biology. Trends Ecol Evol 23:20–25. https://doi.org/10.1016/j.tree.2007.10.003

Aune K, Plumb G (2019) Theodore Roosevelt and Bison restoration of the great plains. History Press, Charleston

Backhouse GN, Clark TW, Reading RP (1995) Reintroductions for recovery of the Eastern Barred Bandicoot Perameles gunnii in Victoria, Australia. In: Serena M (ed) Reintroduction biology of Australian and New Zealand Fauna. Surrey Beatty & Sons, Chipping Norton, pp 209–218

Berger L, Speare R, Daszak P et al (1998) Chytridiomycosis causes amphibian mortality associated with population declines in the rain forests of Australia and Central America. PNAS 95(15):9031–9036

Brown PB, Holdsworth MC, Rounsevell DE (1995) Captive breeding and release as a means of increasing the Orange-bellied Parrot population in the wild. In: Serena M (ed) Reintroduction biology of Australian and New Zealand Fauna. Surrey Beatty & Sons, Chipping Norton, pp 135–141
Bubac CM, Johnson AC, Fox JA, Cullingham CI (2019) Conservation translocations and post-release monitoring: identifying trends in failures, biases, and challenges from around the world. Biol Conserv 238. https://doi.org/10.1016/j.biocon.2019.108239
Cade TMJ, Jones CG (1993) Progress in restoration of the Mauritius Kestrel. Conserv Biol 7:169–175. https://doi.org/10.1046/j.1523-1739.1993.07010169.x
Campbell Grant EH, Miller DAW, Muths E (2020) A synthesis of evidence of drivers of amphibian declines. Herpetologica 76:101–107. https://doi.org/10.1655/0018-0831-76.2.101
Ceballos G, Ehrlich PR, Barnosky AD et al (2015) Accelerated modern human-induced species losses: entering the sixth mass extinction. Sci Adv 1:9–13. https://doi.org/10.1126/sciadv.1400253
Ceballos G, Ehrlich PR, Dirzo R (2017) Biological annihilation via the ongoing sixth mass extinction signaled by vertebrate population losses and declines. Proc Natl Acad Sci USA 114:E6089–E6096. https://doi.org/10.1073/pnas.1704949114
Cowie RH, Bouchet P, Fontaine B (2022) The Sixth Mass Extinction: fact, fiction or speculation? Biol Rev 97:640–663. https://doi.org/10.1111/brv.12816
Crates R, Stojanovic D, Heinsohn R (2023) The phenotypic costs of captivity. Biol Rev 98:434–449. https://doi.org/10.1111/brv.12913
Dobson A, Lyles A (2000) Black-footed ferret recovery. Science 288:985–988. https://doi.org/10.1126/science.288.5468.985
Dodd CHJ, Seigel RA (1991) Relocation, repatriation, and translocation of amphibians and reptiles: are they conservation strategies that work? Herpetologica 47(3):336–350
Ewen JG, Armstrong D, Parker KA, Seddon PJ (eds) (2012) Reintroduction biology: integrating science and management. Wiley-Blackwell, Hoboken
Fischer J, Lindenmayer DB (2000) An assessment of the published results of animal relocations. Biol Conserv 96:1–11. https://doi.org/10.1016/S0006-3207(00)00048-3
Fisher MC, Garner TWJ (2020) Chytrid fungi and global amphibian declines. Nat Rev Microbiol 18:332–343. https://doi.org/10.1038/s41579-020-0335-x
Gascon C, Collins J, Moore R, et al (2007) Amphibian conservation action plan. Proceedings of the IUCN/SSC Amphibian Conservation Summit 2005. Gland
Gascon C, Collins JP, Church DR et al (2012) Scaling a global plan into regional strategies for amphibian conservation. Alytes 29:15–27
Germano JM, Bishop PJ (2009) Suitability of amphibians and reptiles for translocation. Conserv Biol 23:7–15. https://doi.org/10.1111/j.1523-1739.2008.01123.x
Gipps JHW (ed) (1991) Beyond captive breeding: re-introducing endangered mammals to the wild: the proceedings of a symposium held at the Zoological Society of London on 24th and 25th November 1989. Oxford University Press, New York
Griffith B, Scott JM, Carpenter JW, Reed C (1989) Translocation as a species conservation tool: status and strategy. Science 245:477–480. https://doi.org/10.1126/science.245.4917.477
Griffiths RA, Pavajeau L (2008) Captive breeding, reintroduction, and the conservation of amphibians. Conserv Biol 22:852–861. https://doi.org/10.1111/j.1523-1739.2008.00967.x
Gross IP, Wilson AE, Wolak ME (2024) The fitness consequences of wildlife conservation translocations: a meta-analysis. Biol Rev 99:348–371. https://doi.org/10.1111/brv.13025
Harding G, Griffiths RA, Pavajeau L (2016) Developments in amphibian captive breeding and reintroduction programs. Conserv Biol 30:340–349. https://doi.org/10.1111/cobi.12612
Hof C, Araújo MB, Jetz W, Rahbek C (2011) Additive threats from pathogens, climate and land-use change for global amphibian diversity. Nature 480:516–519. https://doi.org/10.1038/nature10650
Hoffmann M, Hilton-Taylor C, Angulo A et al (2010) The impact of conservation on the status of the world's vertebrates. Science 330:1503–1509. https://doi.org/10.1126/science.1194442

Houlahan JE, Findlay CS, Schmidt BR et al (2000) Quantitative evidence for global amphibian population declines. Nature 404:752–755

Hudson M (2016) Conservation management of the mountain chicken frog. Ph.D. thesis. University of Kent

IPBES (2019) Summary for policymakers of the global assessment report on biodiversity and ecosystem services of the Intergovernmental Science-Policy Platform on Biodiversity and Ecosystem Services. Bonn

IUCN/SSC (2013) Guidelines for reintroductions and other conservation translocations. IUCN Species Survival Commitee, Gland

Jachowski D, Millspaugh J, Angermeier P (eds) (2016) Reintroduction of fish and wildlife populations. University of California Press, Oakland

Jenkins CN, Pimm SL, Joppa NJ (2013) Global patterns of terrestrial vertebrate diversity and conservation significance. Proc Natl Acad Sci USA 110: https://doi.org/10.1073/pnas.1302251110

Jiang Z, Yu C, Feng Z et al (2000) Reintroduction and recovery of Pere David's deer in China. Wildl Soc Bull 28:681–687

Jones CG, Heck W, LewisS RE et al (1995) The restoration of the Mauritius Kestrel Falco punctatus population. Ibis (Lond 1859) 137:S173–S180. https://doi.org/10.1111/j.1474-919X.1995.tb08439.x

Jule KR, Leaver LA, Lea SEG (2008) The effects of captive experience on reintroduction survival in carnivores: a review and analysis. Biol Conserv 141:355–363. https://doi.org/10.1016/j.biocon.2007.11.007

Kleiman DG (1989) Reintroduction of captive mammals for conservation. Bioscience 39:152–161. https://doi.org/10.2307/1311025

Lee WG, Jamieson IG (eds) (2001) The Takahe: fifty years of conservation management and research. University of Otago Press, Dunedin

Lewis CHR, Richards-Zawacki CL, Ibáñez R et al (2019) Conserving Panamanian harlequin frogs by integrating captive-breeding and research programs. Biol Conserv 236:180–187. https://doi.org/10.1016/j.biocon.2019.05.029

Longcore JE, Pessier AP, Nichols DK (1999) Batrachochytrium dendrobatidis gen. et sp. nov., a chytrid pathogenic to amphibians. Mycologia 91:219–227

Martel A, Spitzen-Van Der Sluijs A, Blooi M et al (2013) Batrachochytrium salamandrivorans sp. nov. causes lethal chytridiomycosis in amphibians. Proc Natl Acad Sci USA 110:15325–15329. https://doi.org/10.1073/pnas.1307356110

Mendelson JR, Lips KR, Gagliardo RW et al (2006) Confronting amphibian declines and extinctions. Science 313:48. https://doi.org/10.1126/science.1128396

Miller KA, Bell TP, Germano JM (2014) Understanding publication bias in reintroduction biology by assessing translocations of New Zealand's herpetofauna. Conserv Biol 28:1045–1056. https://doi.org/10.1111/cobi.12254

Miskelly CM, Powlesland RG (2013) Conservation translocations of New Zealand birds, 1863–2012. Notornis 60:3–28

Moro D (2003) Translocation of captive-bred dibblers Parantechinus apicalis (Marsupialia: Dasyuridae) to Escape Island, Western Australia. Biol Conserv 111:305–315. https://doi.org/10.1016/S0006-3207(02)00296-3

Morris K, Page M, Kay R et al (2015) Forty years of fauna translocations in Western Australia: lessons learned. In: Armstrong DP, Hayward MW, Moro D, Seddon PJ (eds) . CSIRO Publishing, Advances in reintroduction biology of Australian and New Zealand fauna, pp 217–235

Moseby KE, Read JL, Paton DC et al (2011) Predation determines the outcome of 10 reintroduction attempts in arid South Australia. Biol Conserv 144:2863–2872. https://doi.org/10.1016/j.biocon.2011.08.003

Pimm SL, Jenkins CN, Abell R et al (2014) The biodiversity of species and their rates of extinction, distribution, and protection. Science 344:1246752. https://doi.org/10.1126/science.1246752

Resende PS, Viana-Junior AB, Young RJ, Azevedo CS (2021) What is better for animal conservation translocation programmes: Soft- or hard-release? A phylogenetic meta-analytical approach. J Appl Ecol 58:1122–1132. https://doi.org/10.1111/1365-2664.13873

Scheele BC, Pasmans F, Skerratt LF et al (2019) Amphibian fungal panzootic causes catastrophic and ongoing loss of biodiversity. Science 1463:1459–1463. https://doi.org/10.1126/science.aav0379

Scheele BC, Hollanders M, Hoffmann EP et al (2021) Conservation translocations for amphibian species threatened by chytrid fungus: a review, conceptual framework, and recommendations. Conserv Sci Pract 3:e524

Seal US (1991) Life after extinction. Symp Zool Soc London 62:39–55

Seddon PJ, Armstrong DP, Maloney RF (2007) Developing the science of reintroduction biology. Conserv Biol 21:303–312. https://doi.org/10.1111/j.1523-1739.2006.00627.x

Seddon PJ, Griffiths CJ, Soorae PS, Armstrong DP (2014) Reversing defaunation: restoring species in a changing world. Science 345:406–412. https://doi.org/10.1126/science.1251818

Serena M (ed) (1995) Reintroduction biology of Australian and New Zealand Fauna. Surrey Beatty & Sons, Chipping Norton

Short J (2009) The characteristics and success of vertebrate translocations within Australia

Skerratt LF, Berger L, Speare R et al (2007) Spread of chytridiomycosis has caused the rapid global decline and extinction of frogs. EcoHealth 4:125–134. https://doi.org/10.1007/s10393-007-0093-5

Smith D, Abeli T, Bruns EB et al (2023) Extinct in the wild: the precarious state of the most threatened group of species on Earth. Science 379. https://doi.org/10.1126/science.add2889

Snyder N, Snyder H (2000) The California condor: a saga of natural history and conservation. Academic Press, San Diego

Stanley Price MR (1989) Animal reintroductions: the Arabian Oryx in Oman. Cambridge University Press

Stepkovitch B, Kingsford RT, Moseby KE (2022) A comprehensive review of mammalian carnivore translocations. Mamm Rev 52:554–572. https://doi.org/10.1111/mam.12304

Stuart SN, Chanson J, Cox N et al (2004) Status and trends amphibian declines and extinctions worldwide. Science 306:1783–1786. https://doi.org/10.1126/science.1103538

Tapley B, Bradfield KS, Michaels C, Bungard M (2015) Amphibians and conservation breeding programmes: do all threatened amphibians belong on the ark? Biodivers Conserv 24:2625–2646. https://doi.org/10.1007/s10531-015-0966-9

Trask A, Canessa S, Moehrenschlager A et al (2020) Extinct-in-the-wild species' last stand. Science 369:516. https://doi.org/10.1126/science.abd4560

Wake DB (2012) Facing extinction in real time. Science 335:1052–1053. https://doi.org/10.1126/science.1218364

Wake DB, Vredenburg VT (2008) Are we in the midst of the sixth mass extinction? A view from the world of amphibians. Proc Natl Acad Sci USA 105:11466–11473. https://doi.org/10.1073/pnas.0801921105

Wolf CM, Griffith B, Reed C, Temple SA (1996) Avian and mammalian translocations: update and reanalysis of 1987 survey data. Conserv Biol 10:1142–1154. https://doi.org/10.1046/j.1523-1739.1996.10041142.x

International Concern About the Kihansi Spray Toad

2

Abstract

In early 1999, Friends of the Earth, an environmental NGO, contacted the World Bank and the Norwegian, Swedish, and German international development organizations and called upon these organizations to address the environmental impacts arising from the Lower Kihansi Hydropower Project and the threats that the hydropower project posed to the recently discovered Kihansi Spray Toad. The World Bank agreed to conduct a detailed review of the environmental impacts of the Lower Kihansi Hydropower Project. The Swedish and Norwegian development agencies requested that the Tanzania Electric Supply Company (TANESCO), the project client, conduct a comprehensive environmental and social impact assessment of the project and develop an environmental mitigation plan. The discovery of the Kihansi Spray Toad represented a challenge also for the government of Tanzania because the government had ratified in 1996 the International Convention on Biological Diversity which commits nation states to conserving their biological diversity.

In early November 1999, Friends of the Earth, a US-based non-governmental environmental organization, wrote the directors of the World Bank and the Norwegian, Swedish, and German international development agencies calling upon these organizations to address the environmental impacts of the recently completed Lower Kihansi Hydropower Project. These letters highlighted that the commissioning of the recently completed hydroelectric facility could result in the global extinction of the recently described Kihansi spray toad (*Nectophrynoides asperginis*) and that these development agencies have a responsibility as co-financiers to prevent the extinction of this species.

W. D. Newmark, *The Kihansi Spray Toad* (Nectophrynoides asperginis): *Crisis Biology in the 21st Century*, Ecological Studies 254, https://doi.org/10.1007/978-3-031-89599-9_2

The recently discovered Kihansi Spray Toad, because of its highly restricted geographic range, qualified in 1999 as a Critically Endangered species under IUCN Red List criteria. Additionally, all species within the genus *Nectophrynoides* were listed in 1999 as Appendix 1 species under the Convention on International Trade in Endangered Species of Wild Fauna and Flora (CITES).[1]

The World Bank country director for Tanzania after consulting with the other co-donors and the government of Tanzania replied in March 2000 to the Friends of the Earth stating that the World Bank would conduct a detailed review of the environmental impacts of the hydroelectric facility on the Kihansi Spray Toad and its highly restricted habitat. This letter also stated that as part of this environmental assessment, the World Bank would review all options that have been proposed by the government, the hydropower contractor, and consultants to mitigate the impacts of the hydroelectric facility on the Kihansi Spray Toad and its critical habitat.

2.1 Environmental and Social Assessments of the Lower Kihansi Hydropower Project

The commissioning of the Lower Kihansi Hydropower Project in December 1999 was the culmination of multiple technical, environmental, and social assessments that had been conducted over the previous half century. The earliest study to examine the hydrological potential of the Kihansi River for hydropower production was conducted between 1955 and 1960 by the United Nations Food and Agriculture Organization (FAO) (Otnes 1960) as part of a broader assessment of the hydrological potential of rivers within the Rufiji River Basin. The Rufiji Basin is the largest river basin in Tanzania and East Africa and covers 177,500 km^2 or 20% of the land area of Tanzania. The Kihansi River catchment, 640 km^2, is located along the eastern escarpment of the southern Udzungwa Mountain within the Kilombero subbasin in the Rufiji Basin. The FAO assessment was followed in the early to mid-1980s by a series of studies conducted by Norconsult, a Norwegian engineering firm, and Acres, a Canadian consulting firm examining the hydroelectric potential of the Kihansi River as part of a broader Rufiji Basin Hydropower Master Plan and Power Development Plan 1985–2010. The Rufiji Basin Master Plan identified the Kihansi River as an excellent site for the construction of a hydroelectric dam because of the river's high average base flow (16 m^3/s)—the amount of water that flows into a river from ground water rather than precipitation—and because of the high potential energy of the river due to the large difference in elevation (853 m) between the top and bottom of the Kihansi Gorge (Fig. 2.1).

In 1990, as part of a feasibility study of the Kihansi Hydroelectric Power Development Project Power VI, the Japanese Power Electric Development Company conducted the first environmental assessment of the impact of hydroelectric development on the Kihansi River ecosystem. This environmental assessment stated that

[1] CITES Appendix 1 lists species that are the most endangered and prohibits the international trade in specimens of these species except when the purpose of the import is not commercial.

Fig. 2.1 Kihansi River and Gorge. (Photo credit: TANESCO)

"since rivers in the planned power station site and its vicinity are not used for fishery, agriculture, water-borne traffic, etc., there will be no effects of the construction works at all…" (Mercier 2000). In response to this report, a World Bank supervision mission to the Power VI project stated that descriptions of the flora and fauna data could be improved through a brief site study. As a result, the World Bank commissioned a second environmental assessment that was conducted in 1991 by a Finnish consulting firm Ekono Energy. This environmental assessment concluded that "the [forest] area to be lost [2.5 km^2] is so small that it will not be a serious environmental loss and is most unlikely to threaten extinction of any of the endemic species ... The loss of habitat is a small price to pay for the economic benefits of power generation" (Ekono Energy 1991). A subsequent internal World Bank environmental assessment of the Power VI project conducted in 1992 concurred with these earlier environmental assessments, and a World Bank Staff Appraisal Report released in

April 1993 concluded that "The run of the river project is not expected to have a serious downstream impact on wildlife or fisheries. The expected [impacts] on flora and fauna should be minimal and temporary" (Mercier 2000).

Based on the World Bank's internal environmental and social assessment, the Board of Directors of the World Bank approved funding in May 1993 for the engineering and administrative works for the Lower Kihansi Hydropower Project. A condition, however, for the World Bank's bank approval was that the Tanzania Electric Supply Company (TANESCO)—the project client—was required to implement an Environmental Management Plan to address all environmental and social impacts arising from the project (Gerstle et al. 1997), which in 1993 were thought to be issues related predominantly to adverse impacts in the inundation area of the dam and to communities surrounding the project area.

In July 1994 the World Bank's Board of Directors approved preparatory works for the construction of infrastructure including staff housing, roads, an airstrip, a diesel generating station (for purposes of construction power), water treatment plants, and a health clinic. In November 1994 the Board of Directors approved construction of the main civil works including the dam and power station. The total project financing for the project was $275 million which included co-financing from Norway, Sweden, Germany, United Kingdom, and the European Investment Bank.

The Lower Kihansi Hydropower Project was designed as a run-of-the-river facility that incorporated a 25 m high dam and 26 ha reservoir (Fig. 2.2) and took

Fig. 2.2 The Kihansi hydropower dam and reservoir are located immediately above the Kihansi Gorge

advantage of the 850 m change in elevation over 3 km between the top and bottom of the Kihansi Gorge. Water from the dam would be diverted into a vertical tunnel which would lead to a power generating station at the base of the Kihansi Gorge and then returned to the river 6 km downstream of the dam. The Lower Kihansi Hydropower Plant, which has a 180 MW capacity, was originally designed to provide electricity during peak demand in the evening over a 4–5-hour period. However, since its commissioning in December 1999, the Lower Kihansi Hydropower Plant has been operated largely to meet base load demand because the Lower Kihansi Hydropower Plant is the most cost-effective electrical generating facility in Tanzania.

2.2 Concerns of Co-financing Partners

Unfortunately, none of the environmental assessments that were conducted prior to the World Bank's Board of Directors approval for the dam had examined potential environment impacts below the dam. Because of this and other shortcomings, a precondition of two of the co-financing partners, the Swedish International Development Cooperation Agency (SIDA) and the Norwegian Agency for Development Cooperation (NORAD), was that TANESCO should conduct a comprehensive environmental and social impact assessment of the project and should also develop an environmental mitigation plan. SIDA was particularly concerned that a comprehensive environmental impact assessment had not been completed prior to the initiation of the project, and in June 1995, SIDA in a letter to TANESCO stated ".... of special importance is that TANESCO will assemble data for clarification of the relation between the minimum flow and the ecosystem downstream of the dam and that monitoring measures like establishment of bypass flow requirements, shall be taken to minimize effects on the flora and fauna in the project area" (Mercier 2000).

In December 1995, NORPLAN, the project contractor for the Lower Kihansi hydropower facility, noted in their finalized environmental assessment that contrary to earlier views the diversion of water for the hydroelectric scheme would very likely have environmental impacts on the vegetation on both sides of the 6 km stretch of the river which passed through the Gorge. This assessment also noted that the fine spray from multiple waterfalls within the gorge was important for maintaining the microclimate and spray vegetation habitats within the Gorge ecosystem and thus maintaining a minimum bypass flow through the gorge was required. Lastly the assessment noted that "Eastern Arc endemic amphibians and reptiles would probably undergo local extinction" (Mercier 2000).

The importance of maintaining a minimum bypass flow through the Gorge was reiterated in early 1996 in an internal environmental assessment conducted by KfW, one of the co-financing partners. In 1996, TANESCO agreed to establish with support from SIDA, NORAD, and KfW the Lower Kihansi Long Term Environmental Monitoring Program to address the environmental impacts identified in the 1995 environmental assessment. Under this environmental monitoring program, an inventory of plants, birds, insects, small mammals, fish, reptiles, and amphibians was conducted. During this inventory not only was the Kihansi Spray Toad discovered but also four new endemic plant species, *Coffea kihansiensis*, *Kihansia lovettii*, *Kupea jonii*, and *Stenandrium grandiflorum* (Vollesen 2000; Cheek 2003; Davis and Mvungi 2004).

2.3 Contributory Factors to Misidentification of Environmental Impacts

The environmental impacts arising from the Lower Kihansi Hydroelectric Project were a result of both omissions and errors on the part of the World Bank and co-financing partners. The most serious shortcoming was World Bank's Board of Directors approval of the Lower Kihansi Hydropower Project prior to the completion of a comprehensive environmental and social impact assessment including a detailed inventory of plant and animal species occurring downstream of the proposed dam site. This lack of a detailed inventory of plant and animal species within the Kihansi Gorge was in retrospect not surprising given that the only on-site environmental assessment conducted prior to 1995 was based on 2 days of fieldwork within the Kihansi project area and the resulting species checklists had only focused on plants, birds, and mammals (Ekono Energy 1991). Amphibians and reptiles were not included in this compilation nor discussed in this brief environmental assessment. Additionally, all of the field work that was conducted in the Kihansi Gorge prior to 1995 was restricted to the mouth of the gorge due very likely to the difficulty of accessing the upper regions of the gorge in the early 1990s.

Prior to the commissioning of the dam, spray generated by the largest waterfalls extended up to 100–200 m away from the river. According to Kim Howell, who discovered the Kihansi Spray Toad in 1996, entering the spray wetlands adjacent to the largest waterfalls in the Kihansi Gorge prior to the commissioning of the dam was "like working next to an open fire hydrant." Without rain gear, it was only possible to work in the spray wetlands for short periods of time.

A second major error on the part of the World Bank and co-financing partners was that the hydropower facility was constructed in the Eastern Arc Mountains, a global biodiversity hotspot for threatened and endemic plant and animal species. Although the Eastern Arc forests had been described in 1990 as 1 of 12 global tropical biodiversity hotspots (Myers 1990), none of the environmental assessments that had been conducted prior to the approval of the Lower Kihansi Hydropower Project had noted this.

2.4 The Eastern Arc Mountains

The Eastern Arc Mountains are a group of isolated mountains stretching from southeast Kenya through south central Tanzania (Lovett and Wasser 1993; Newmark 2002). They consist of the Taita Hills in Kenya and the Pare, Usambara, Nguru, Ukaguru, Rubeho, Uluguru, Mahenge, Malundwe, and Udzungwa Mountains in

Tanzania (Fig. 2.3). Their age, geologic origin, and proximity to the Indian Ocean are what separate them from other highland regions in East Africa (Lovett and Wasser 1993). These same features have also contributed to their very diverse and unique biota, which is quite distinct from the biota found in adjacent savannah and woodland habitats in East Africa. The Kihansi River and Gorge in the Udzungwa Mountains are located along the very southern edge of the Eastern Arc Mountains.

The total area of natural forest in the Eastern Arc Mountains is 4059 km^2 (Newmark and McNeally 2018) or 0.5% of the total land area of Tanzania. The Udzungwa Mountains, which contains the entire catchment of the Kihansi River, are the largest of the 13 Eastern Arc Mountains and contain one-half (48%) of all remaining forest in the Eastern Arc Mountains. However, the remaining natural forest in the Udzungwa Mountains as in the other Eastern Arc Mountains is highly

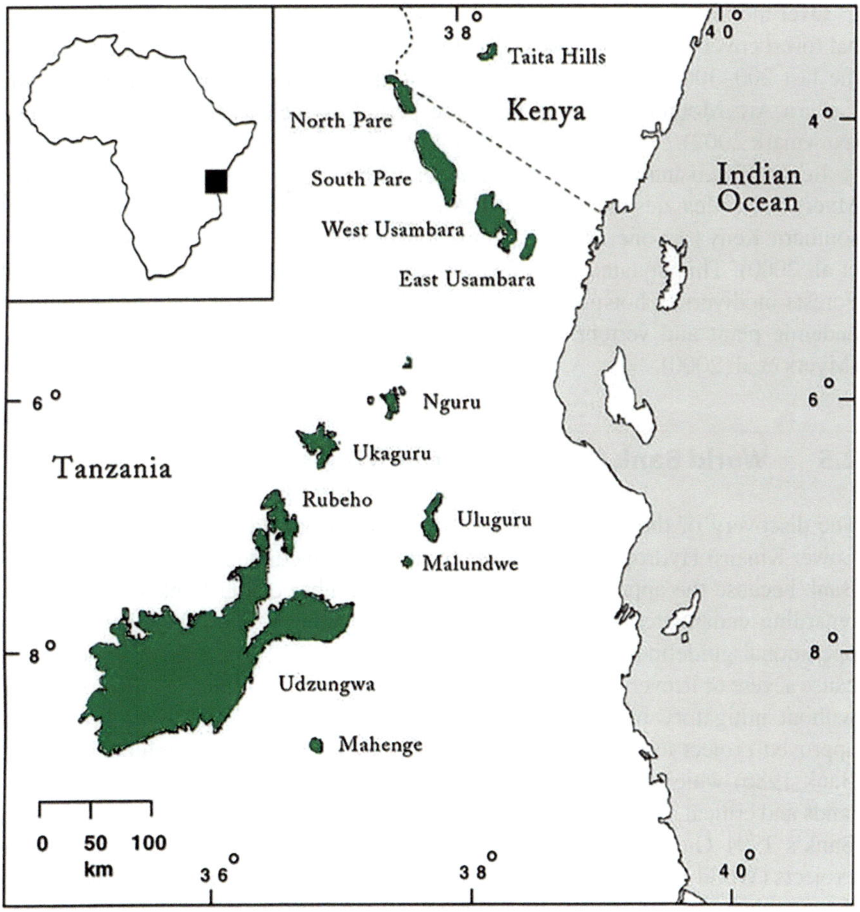

Fig. 2.3 Location of the Eastern Arc Mountains in Kenya and Tanzania

fragmented comprising of 67 fragments >10 ha in size with a median patch size of 55 ha (Newmark and McNeally 2018).

For their size, the Eastern Arc forests have the highest number of plant and animal species of any region in Tanzania. The Eastern Arc forests contain approximately 27% of all plant species, 63% of all linyphiid spider species, 43% of all butterfly species, 33% of all amphibian species, 37% of all reptile species, 37% of all bird species, and 34% of all mammal species that occur in Tanzania (Newmark 2002).

Yet what is particularly striking about the Eastern Arc Mountains is the unusually large numbers of endemic species (Myers et al. 2000; Burgess et al. 2007). Of the known species occurring within the Eastern Arc forests, approximately 23% of montane plant species, 82% of linyphiid spider species, 39% of the butterflies species, 63% of the forest-dependent amphibian species, 68% of the forest-dependent reptile species, 3% of the bird species, and 6% of the mammal species are endemic to these forests (Newmark 2002).

Over the last 2000 years, the Eastern Arc Mountains have lost 77% of the original forest cover (Newmark and McNeally 2018). Most of this loss has occurred over the last 200–300 years due to the dramatic increase in human population in the Eastern Arc Mountains during this time period as well as technological change (Newmark 2002).

In an updated analysis of global biodiversity hotspots published in 2000, Norman Myers and colleagues identified the Eastern Arc and coastal forests of Tanzania and southern Kenya as one of the 25 most threatened ecosystems worldwide (Myers et al. 2000). This updated analysis also reported that the Eastern Arc and Coastal Forests biodiversity hotspot contained, due to their small size, the highest ratio of endemic plant and vertebrate species to area of any global biodiversity hotspot (Myers et al. 2000).

2.5 World Bank Operational Guidelines

The discovery of the Kihansi Spray Toad after the World Bank's approval of the Lower Kihansi Hydropower Project presented an immediate challenge to the World Bank because the approved project violated a number of the Bank's own policies regarding endangered species and their critical habitat. The 1984 World Bank's operational guidelines stated that the World Bank "will not finance projects that cause severe or irreversible environmental degradation, including species extinction without mitigatory measures acceptable to the Bank" (World Bank 1984). The approved project also violated the 1986 World Bank policy on wildlands (World Bank 1986) which focused on conserving biodiversity through conserving wildlands and critical species habitat. And lastly, the approved project ignored the World Bank's 1991 Guidelines for Environmental Assessment of Energy and Industry Projects (World Bank 1991). These guidelines stated that "the area of influence of a dam and its reservoir extends from the upper limits of the reservoir to as far downstream as the estuarine, coastal and offshore zones… and the greatest impacts result from the impoundment of water, flooding of land to form the reservoir, and alteration of water flow downstream."

2.6 International Convention on Biological Diversity

The discovery the Kihansi Spray Toad presented not only a serious challenge for the World Bank and other international donors but also the Tanzanian government. In March 1996, the United Republic of Tanzania ratified the International Convention on Biological Diversity. This convention, which is a multi-lateral treaty that has been signed and ratified by more than 190 nations, commits nation states to conserving their biological diversity. The specific goals of this convention are (1) the conservation of biological diversity; (2) the sustainable use of its components; and (3) the fair and equitable sharing of benefits arising from genetic resources.

The signed International Convention on Biological Diversity made ignoring the environmental impacts of the Lower Kihansi Hydropower Project on the Kihansi Spray Toad and its critical habitat extremely difficult for both the donor community and the government of Tanzania. Thus, the World Bank's operational guidelines, bilateral donors' concerns, and the International Convention on Biological Diversity were important drivers behind the development of a series of projects that were developed over the following two decades to mitigate the environmental impacts arising from the Lower Kihansi Hydropower Project.

References

Burgess ND, Butynski TM, Cordeiro NJ et al (2007) The biological importance of the Eastern Arc Mountains of Tanzania and Kenya. Biol Conserv 134:209–231. https://doi.org/10.1016/j.biocon.2006.08.015

Cheek M (2003) Kupeaeae, a new tribe of Triuridaceae from Africa. Kew Bull 58:939–949

Davis AP, Mvungi EF (2004) Two new and endangered species of Coffea (Rubiaceae) from the Eastern Arc Mountains (Tanzania) and notes on associated conservation issues. Bot J Linn Soc 146:237–245. https://doi.org/10.1111/j.1095-8339.2004.00328.x

Ekono Energy (1991) Kihansi hydroelectric project environmental assessment EP-13624Q-01. Nordic Development Fund, Helsinki

Gerstle JH, Mhaville SL, Lindemark J (1997) Environmental aspects of the lower Kihansi hydropower project, Tanzania. In: Broch E, Lysne DK, Flatabo N, Helland-Hasen E (eds) Hydropower '97. Proceedings of the 3rd international conference on hydropower Trondheim/Norway 30 June – 2 July 1997. A.A. Balkema/R, Rotterdam, pp 23–31

Lovett JC, Wasser SK (eds) (1993) Biogeography and ecology of the rain forests of eastern Africa. Cambridge University Press, Cambridge

Mercier JR (2000) Lower Kihansi hydropower project (LKHP)—Tanzania power VI: chronology of biodiversity, water rights and mitigative measures. Internal non-official World Bank report

Myers N (1990) The biodiversity challenge: expanded hot-spots analysis. Environmentalist 10:243–256. https://doi.org/10.1007/BF02239720

Myers N, Mittermeier RA, Mittermeier CG et al (2000) Biodiversity hotspots for conservation priorities. Nature 403:853–858. https://doi.org/10.1038/35002501

Newmark WD (2002) Conserving biodiversity in East African forests: a study of the Eastern Arc Mountains. Springer, Berlin

Newmark WD, McNeally PB (2018) Impact of habitat fragmentation on the spatial structure of the eastern arc forests in East Africa: implications for biodiversity conservation. Biodivers Conserv 27:1387–1402. https://doi.org/10.1007/s10531-018-1498-x

Otnes J (1960) The Rufiji Basin Tanganyika FAO report to the Government of Tanganyika on the preliminary reconnaissance survey of the Rufiji Basin. FAO, Rome

Vollesen K (2000) Two new Tanzanian Acanthaceae. Kew Bull 55:965–969

World Bank (1984) Operation manual 2.35 environmental aspects of bank work

World Bank (1986) Operational policy note 11.02 wildlands: their protection and management in economic development

World Bank (1991) Environmental assessment sourcebook, Guidelines for environmental assessment of energy and industry projects, vol III. World Bank, Washington, D.C.

Discovery of the Kihansi Spray Toad

3

Abstract

In December 1996, Kim Howell a herpetologist from the University of Dar es Salaam in Tanzania and two colleagues discovered the Kihansi Spray Toad during an ecological survey of the Kihansi Gorge. A formal description of the Kihansi Spray Toad was published in 1998, and in 1999 the Kihansi Spray Toad was listed as Critically Endangered. The Kihansi Spray Toad is a habitat specialist and was restricted at the time of its discovery to five spray wetlands that were maintained by spray from natural waterfalls along the Kihansi River. The total area of the spray wetlands is 0.49 ha, and thus the geographic range of the Kihansi Spray Toad was one of the smallest for any tetrapod worldwide. The Kihansi Spray Toad is a generalist feeder, ovoviviparous—the toadlets when born are 2 mm in length—and its call notes extend into the ultrasonic frequency, which has been suggested to be an adaptation to the high background noise level of the spray toad's habitat.

In December 1996, the first ecological surveys were conducted within the Kihansi Gorge under the NORPLAN-supported Lower Kihansi Long Term Environmental Monitoring Program. As part of these ecological surveys, Kim Howell an American professor in the Department of Zoology and Wildlife Conservation at the University of Dar es Salaam and who had resided in Tanzania for more than 30 years (Fig. 3.1), along with Peter Hawkes an entomologist from South Africa and David Moyer an ornithologist from Tanzania, collected several specimens of an unknown anuran in the Upper Spray Wetland in the Kihansi Gorge. Kim Howell immediately recognized that this unknown anuran was almost certainly a new species. In late 1997, Kim Howell returned to the Kihansi Gorge to collect type specimens and to assess

W. D. Newmark, *The Kihansi Spray Toad* (Nectophrynoides asperginis): *Crisis Biology in the 21st Century*, Ecological Studies 254, https://doi.org/10.1007/978-3-031-89599-9_3

Fig. 3.1 Kim Howell, a herpetologist at the University of Dar es Salaam, discovered the Kihansi Spray Toad in December 1996. (Photo credit: Abubakar S. Ringim)

the abundance of this new anuran in the Kihansi Gorge. The collected specimens were deposited in the British Museum in London. A year later Kim Howell and colleagues formally described this new anuran and was named the Kihansi Spray Toad (*Nectophrynoides asperginis*) (Poynton et al. 1998). *Aspergo* in Latin means spray.

3.1 Species Description

The Kihansi Spray Toad is yellow to mustard in color (Fig. 3.2) with dark dorsal markings and is sexually dimorphic (Poynton et al. 1998). The males have dark inguinal patches and are smaller in size and body mass than the females. The mean body length of an adult male is 17.6 mm, and that of an adult female is 21.8 mm (Channing et al. 2006). Mean body mass across all age classes (juvenile, subadult, adult) is 0.3 g. Body mass of adults ranges from 0.45 to 0.66 g (Lee et al. 2006) with gravid females (Fig. 3.3) weighing 0.8–0.9 g (Channing et al. 2006; Lee et al. 2006).

3.2 Ecological Specialization

The Kihansi Spray Toad is an ecological specialist. The Kihansi Spray Toad has extensive webbing on both fingers and toes that allow them to easily climb wet vertical rock facers. Channing et al. (2006) state that "the thin skin restricts the toad to cool, humid areas, while the small size and climbing ability allow it to live on the rock faces within the direct spray zone of the various falls."

Fig. 3.2 Kihansi Spray Toad. (Photo credit: Timothy Herman)

Fig. 3.3 A gravid female Kihansi Spray Toad. (Photo credit: TANESCO)

3.3 Habitat

The Kihansi Spray Toad was restricted in 1996, the year of its discovery, to five spray wetlands located along the Kihansi River within the Kihansi Gorge (Fig. 3.4) (Poynton et al. 1998). The Kihansi Gorge, through which the Kihansi River flows,

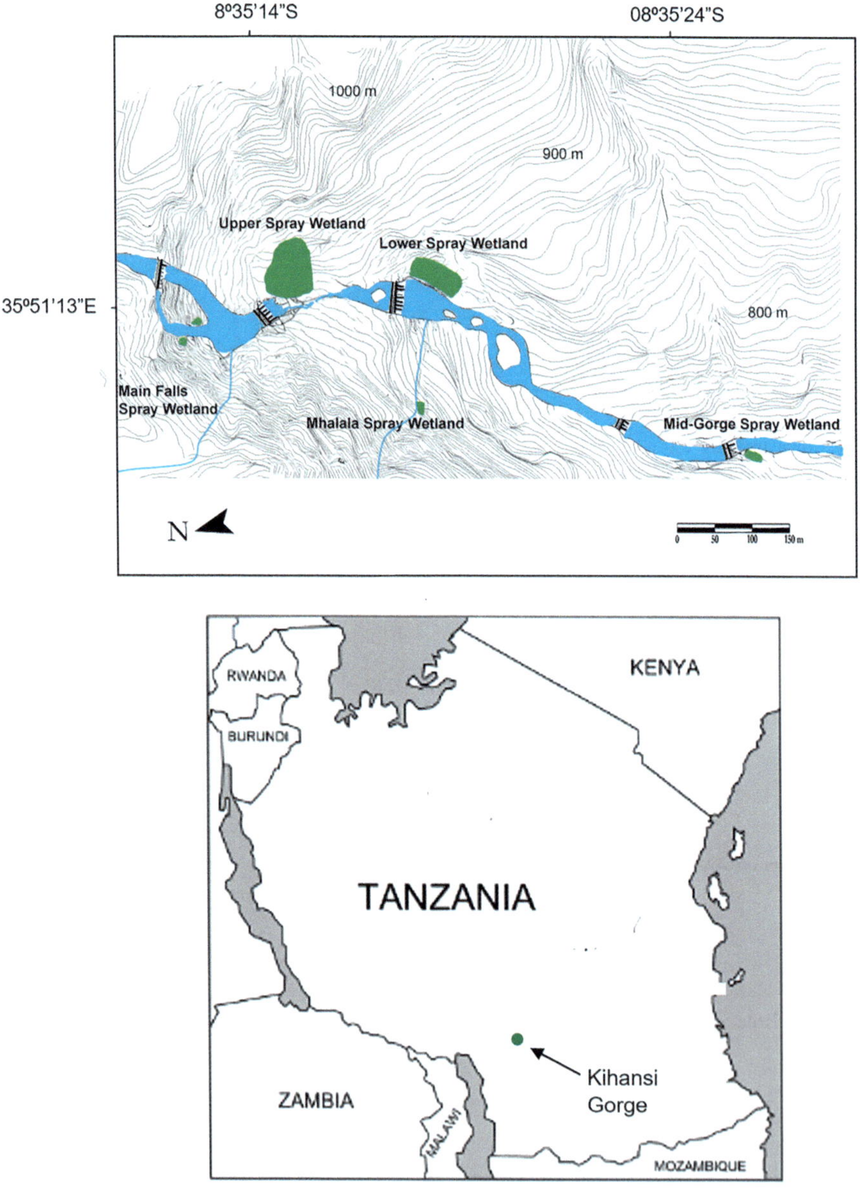

Fig. 3.4 Location of the Main Falls, Upper, Lower, Mhalala, and Mid-Gorge Spray Wetlands in relation to the Kihansi River within the Kihansi Gorge. (After NORPLAN (2002a))

is approximately 200 ha in area (~4.22 km × ~0.47 km). The Kihansi River as it flows through the gorge plunges over a series of waterfalls and rapids (Fig. 3.1). The Main Falls at the head of the steep-sided gorge is 70 m in height, and the Lower Falls, the second largest waterfall in the Kihansi Gorge, is 15 m in height. There are also a series of smaller waterfalls ranging in height between 5 and 10 m along upper and middle regions of the Kihansi Gorge (Fig. 3.4).

Four of the five spray wetlands (Main Falls, Upper, Lower, and Mid-Gorge) are located adjacent to waterfalls along the Kihansi River (Fig. 3.4). The fifth spray wetland is located 180 m upslope from the Kihansi River adjacent to a waterfall along the Mhalala stream, a tributary of the Kihansi River.

The spray wetlands adjacent to the Kihansi River and the Mhalala Stream were created and maintained by spray generated by the adjacent waterfalls. The spray wetland vegetation is comprised largely of moss (*Selaginella kraussiana*), grasses (*Panicum* spp.), forbs (*Alchemilla kiwuensis*, *Begonia oxyloba*, *Brillaintaisia madagascariensis*, *Drymaria cordata*, *Impatiens* spp., *Rumex abyssinicus*, *Streptocarpus buchananii*), and ferns (*Tectaria gemmifera*) (Poynton et al. 1998; Vandvik et al. 2014) (Figs. 3.5, 3.6, 3.7, and 3.8).

Inspection of recent high-resolution Google Earth imagery and detailed maps of the spray wetlands and adjacent rocky shoreline produced under the Immediate Rescue Emergency Measures Project (IREM) reveal that the original area of the five spray wetlands was approximately 4900 m^2 (see below). However, prior to the commissioning of the hydroelectric facility, an area as large as ~15,000 m^2 within the Gorge very likely received spray generated by the numerous waterfalls and rapids along the Kihansi River (Channing et al. 2006). The spray that was generated by the

Fig. 3.5 The Main Falls Spray Wetland, smallest of the five spray wetlands, is located above the Main Falls plunge pool along the western branch of the Kihansi River. (Photo credit: TANESCO)

Fig. 3.6 The Upper Spray Wetland is the largest of the five spray wetlands and is located along the eastern edge of Kihansi River downstream of the Main Falls. This photograph was taken prior to the installation of the spray irrigation systems. (Photo credit: TANESCO)

Fig. 3.7 The Lower Spray Wetland is the second largest spray wetland in the Kihansi Gorge and is located along the eastern edge of the Kihansi River and downstream from the Upper Spray Wetland

Fig. 3.8 The Mhalala Spray Wetland, the second smallest of the five spray wetlands, is located 180 m upslope and west of the Kihansi River along the Mhalala Stream a tributary of the Kihansi River and was the only spray wetland that was not directly impacted by the commissioning of the hydroelectric facility

waterfalls along the Kihansi River prior to the commissioning of the hydroelectric facility extended into the forest bordering spray wetlands and covered all of the large boulders and rock bordering the river and spray wetlands. Rock faces along the Kihansi River (Fig. 2.1) between the Main Falls and the Mid-Gorge Spray Wetland also very likely provided historically additional habitat for the Kihansi Spray Toad (Channing et al. 2006) (Fig. 3.9).

3.4 Diet

The Kihansi Spray Toad is a generalist feeder, and its diet consists of a wide variety of small arthropods ranging in size from 0.3 mm to 10.0 mm (Channing et al. 2006). In surveys conducted between 2000 and 2002 of the stomach contents of the Kihansi Spray Toad, Diptera comprised 60.3% of stomach contents by number, Homoptera 9.7%, Hymenoptera 7.5%, and Acarina 11.4% (NORPLAN 2002b). The two most common invertebrate species that were identified in the stomach contents prior to the commissioning of the dam were a leafhopper *Afrosteles distans* and an ensign scale *Ortheziola* sp. and comprised 9.2% of the identified stomach contents by number. Both of these invertebrate species are abundant in the spray wetlands (NORPLAN 2002b).

Fig. 3.9 The Mid-Gorge Spray Wetland, the third largest and lowest elevation spray wetland in the Kihansi Gorge, is located along the western edge of the Kihansi river. (Photo credit: TANESCO)

3.5 Breeding and Reproduction

The Kihansi Spray Toad is ovoviviparous. Fertilization is internal and the spray toad gives live birth to babies. The baby toadlets when born are 2 mm in length. Observed clutch sizes in the wild ranged from 5 to 13 (Channing et al. 2006). In captivity, mean (± 1 SD) clutch size is 9 ± 5, and the maximum recorded clutch size is 22 (Lee et al. 2006).

Ovoviviparity, although uncommon, is not unique to the Kihansi Spray Toad. Other species in the genus *Nectophrynoides* are known to be ovoviviparous (Wake 1980; Menegon et al. 2004; Channing et al. 2006). Of the 11 described species of *Nectophrynoides* in Tanzania, at least 5 species are known to be ovoviviparous, and all of these species occur in montane habitats above 400 m (Menegon et al. 2004; Channing and Howell 2006; Channing et al. 2006).

In the wild, breeding in the Kihansi Spray Toad was recorded as highly seasonal with peak amplexus occurring in December through January with high numbers of juveniles observed in March through April and subadults between May and July (Channing et al. 2006). Amplexus is prolonged and occurs frequently on rock faces (Channing et al. 2006). The female presses her abdomen to the rock face, and the male adpresses his cloaca to the female. Males have been seen in amplexus at

7 months, and females have had visible follicle development by 8.5 months (Lee et al. 2006). The gestation period is 30–60 days (Lee et al. 2006).

In the Toledo zoo, peak number of births between 2000 and 2019 occurred in June, a temporal shift of 2–3 months later than that observed in the wild (Odum 2018). The maximum life span of the Kihansi Spray Toad in captivity is 5 years (CBSG 2008).

3.6 Spatial Distribution

In the wild, the spatial distribution of the Kihansi Spray Toad within the spray wetlands varied by age class. Juveniles were entirely restricted to the ground layer within the spray wetlands, while subadults were found in the lower layers of the vegetation and occasionally on rock faces. Adults occurred throughout the vegetation in the spray wetlands as well as on rock faces (Channing et al. 2006).

The spatial distribution of the Kihansi Spray Toad within the spray wetlands also varied diurnally and seasonally. At night, the subadult and adult spray toads move from lower layers in the vegetation to the top of the vegetation and onto rock faces with peak number of toads on rock faces occurring at 24:00 hr. During the dry season (July–February), the Kihansi Spray Toad would move closer to the river forming small aggregations on wet rock overhangs and vertical surfaces in the wetland. During the wet season (March–June), the spray toad would disperse throughout the spray wetland (Channing et al. 2006).

3.7 Acoustic Signaling

Kihansi Spray Toad males use calls during male-to-male agonistic signaling and during courtship. Males tend to congregate and form choruses while concealed in vegetation. Calls directed at males are initiated when another male is within 200 mm (Channing et al. 2006). Calls directed at males are shorter in duration, have longer internote intervals, longer in interbout intervals, and are lower in frequency than calls directed at females (Arch et al. 2011). One of the more notable aspects about the call notes of the Kihansi Spray Toad is that they extend into the ultrasonic frequency (Arch et al. 2011). This has been described in only two other anuran species. Arch et al. (2011) hypothesize this is an adaptation to the high background noise level of the Kihansi Spray Toad's habitat due to constant splashing water and high wind. Equally fascinating is that the Kihansi Spray Toad is earless. While the KST does possess a typical anuran inner ear, this inner ear lacks a tympanic membrane and air-filled middle cavity (Arch et al. 2011). Earless anurans are sensitive to airborne sound but must rely on extratympanic sound transmission routes. Arch et al. (2011) suggest that the vocal behavior for the Kihansi Spray Toad is, thus, best suited for short-range intraspecific communication.

3.8 Geographic Range

The Kihansi Spray at the time of its discovery was restricted to five wetlands located along the Kihansi River and along one tributary in the Kihansi Gorge. One individual was collected in December 1996 in a very small low-elevation wetland (~395 m) along the Kihansi River (Poynton et al. 1998). However, since 1996 no other individuals have been observed at this low-elevation site.

Based on an inspection of high-resolution Google Earth imagery and detailed maps of the spray wetlands and adjacent rocky shoreline produced under the IREM project, the total area of the five wetlands located within the Kihansi Gorge is 4907 m^2 (Table 5.1). Thus, the historic geographic range of the Kihansi Spray Toad, ~0.49 ha, is among the smallest for any tetrapod worldwide (Newmark 2002; Ficetola et al. 2015; Meiri et al. 2018).

References

Arch VS, Richards-Zawaki CL, Feng AS (2011) Acoustic communication in the Kihansi Spray Toad (Nectophrynoides asperginis): insights from a captive population. J Herpetol 45:45–49. https://doi.org/10.1670/10-084.1

CBSG (2008) Population and habitat viability assessment for the Kihansi Spray Toad, Apple Valley

Channing A, Howell KM (2006) Amphibians of East Africa. Cornell University Press, Ithaca

Channing A, Finlow-Bates KS, Haarklau SE, Hawkes PG (2006) The biology and recent history of the critically endangered Kihansi spray toad Nectophrynoides asperginis in Tanzania. J East African Nat Hist 95:117–138. https://doi.org/10.2982/0012-8317(2006)95[117:tbarho]2.0.co;2

Ficetola GF, Rondinini C, Bonardi A et al (2015) Habitat availability for amphibians and extinction threat: a global analysis. Divers Distrib 21:302–311. https://doi.org/10.1111/ddi.12296

Lee S, Zippel K, Ramos L, Searle J (2006) Captive-breeding programme for the Kihansi spray toad Nectophrynoides asperginis at the Wildlife Conservation Society, Bronx, New York. Int Zoo Yearb 40:241–253. https://doi.org/10.1111/j.1748-1090.2006.00241.x

Meiri S, Bauer AM, Allison A et al (2018) Extinct, obscure or imaginary: the lizard species with the smallest ranges. Divers Distrib 24:262–273. https://doi.org/10.1111/ddi.12678

Menegon M, Salvidio S, Loader SP (2004) Five new species of Nectophrynoides noble 1926 (Amphibia Anura Bufonidae) from the Eastern Arc Mountains, Tanzania. Trop Zool 17:97–121. https://doi.org/10.1080/03946975.2004.10531201

Newmark WD (2002) Conserving biodiversity in East African forests: a study of the Eastern Arc Mountains. Springer, Berlin

NORPLAN (2002a) Lower Kihansi hydropower project immediate rescue and emergency measures. Final report. Report produced for Tanzania Electric Supply Company Ltd (TANESCO), Dar es Salaam, Tanzania

NORPLAN (2002b) Lower Kihansi hydropower project immediate rescue and emergency measures. Final specialist report: amphibian studies. Report produced for Tanzania Electric Supply Company Ltd (TANESCO), Dar es Salaam, Tanzania

Odum A (2018) Kihansi spray toad captive population. Excel database. Toledo Zoo, Toledo

Poynton JC, Howell KM, Clarke BT, Lovett JC (1998) A critically endangered new species of Nectophrynoides (Anura: Bufonidae) from the Kihansi gorge, Udzungwa Mountains, Tanzania. African J Herpetol 47:59–67

Vandvik V, Maren IE, Ndangalasi HJ et al (2014) Back to Africa: monitoring post-hydropower restoration to facilitate reintroduction of an extinct-in-the-wild amphibian. Ecosphere 5. https://doi.org/10.1890/ES14-00093.1

Wake MH (1980) The reproductive biology of Nectophrynoides malcolmi (Amphibia: Bufonidae), with comments on the evolution of reproductive modes in the genus Nectophrynoides. Copeia 1980:193–209

Proposed Mitigation Options to Conserve the Kihansi Spray Toad and Its Habitat, 1995–1999

4

Abstract

NORPLAN, the project contractor for the Lower Kihansi Hydropower Project, proposed five mitigation options between 1995 and 1999 to conserve the Kihansi Spray Toad and its critical habitat. These included (1) establishment of a minimum bypass flow for the Kihansi River; (2) creation of an artificial spray system to maintain the spray wetland vegetation; (3) intermittent bypass flow releases from the dam to maintain the spray wetland vegetation; (4) establishment of a captive breeding population of the Kihansi Spray Toad; and (5) conduct an expanded search for the Kihansi Spray Toad and spray wetland habitat outside of the Kihansi Gorge and evaluate the feasibility of translocation of the Kihansi Spray Toad to another site in the Udzungwa Mountains.

In response to concerns raised by bi-lateral donors about the importance of assessing minimum bypass flows on plant and animal communities downstream of the dam, NORPLAN conducted in mid-1995 a new environmental assessment of the Lower Kihansi Hydropower Project. In this this assessment, NORPLAN concluded that contrary to conclusions of earlier environmental assessments, the diversion of water for the hydroelectric facility would very likely have significant environmental impacts on the vegetation within the Gorge and that maintaining the spray regime within the gorge would be important to conserving unusual plant communities within the Gorge.

Between 1995 and 1999, NORPLAN proposed under the Lower Kihansi Long Term Environmental Monitoring Program five mitigation options to address the environmental impacts arising from the Lower Kihansi Hydropower Project. These options included:

W. D. Newmark, *The Kihansi Spray Toad* (Nectophrynoides asperginis): *Crisis Biology in the 21st Century*, Ecological Studies 254, https://doi.org/10.1007/978-3-031-89599-9_4

1. Establishment of a minimum bypass flow for the Kihansi River
2. Creation of an artificial spray system to maintain the spray wetland vegetation
3. Intermittent bypass flow releases from the dam to maintain the spray wetland vegetation
4. Establishment of a captive breeding population of the Kihansi Spray Toad
5. Conduct an expanded search for the Kihansi Spray Toad and its spray wetland habitat outside of the Kihansi Gorge and evaluate the feasibility of translocation of the Kihansi Spray Toad to another site in the Udzungwa Mountains

4.1 Establishment of a Minimum Bypass Flow for the Kihansi River

In 1996, the Rufiji Basin Water Office (RBWO) issued TANESCO a provisional water right to abstract water from the Kihansi River. This provisional water right stated that a minimum bypass flow of 7.0 m^3/s of water should be maintained in the Kihansi River and that all water abstracted from the Kihansi River should be returned undiminished and unpolluted to the river downstream from the power plant. This minimum bypass flow was based on the minimum recorded historic flow of 7.7 m^3/s derived from 30-day low flow data between 1975 and 1998. The average historic flow in the Kihansi River over this same time period was 16.3 m^3/s (Hirji 2005).

TANESCO, however, immediately contested the provincial water right and stated that maintaining a minimum bypass flow of 7.0 m^3/s would make the Lower Kihansi hydropower facility economically infeasible. According to TANESCO's calculations that were conducted by NORPLAN under the Immediate Rescue and Emergency Measures Project (IREM), the net present value in 1999 of having to purchase alternative thermal energy (diesel) as a result of a constant 8.0 m^3/s bypass flow was 230 million USD per year, while the net present value for a constant 1.5m^3/s bypass was 50 million USD per year (NORPLAN 2002). Thus, TANESCO expressed a strong preference for the smallest bypass option (1.5 m^3/s) that NORPLAN modeled.

In March 2000, the Rufiji Basin Water Office (RBWO) carried out an inspection of the Lower Kihansi hydropower facility. The RBWO inspection revealed that the provisional water right had been violated and that TANESCO had installed during the construction of the hydroelectric dam a bypass structure that was capable of releasing 1.51–1.89 m^3/s.

In May 2000, the Division of Environment in the Vice-President's Office recommended that TANESCO maintain a minimum bypass flow of 7.7 m^3/s for a period of 1–2 years in order to ensure the survival of the Kihansi Spray Toad species pending further studies. When the Lower Kihansi hydropower facility was inaugurated by President of Tanzania in July 2000, the Lower Kihansi hydropower facility was operating without a final water right, and a minimum bypass flow in the Kihansi River of 1.51–1.89 m^3/s was being maintained by the bypass structure that TANESCO had installed during the construction of the dam.

Thus, defining a final water right for the operation of the Lower Kihansi hydropower facility became an important objective of the first World Bank supported follow-on project to the Lower Kihansi Hydropower Project.

4.2 Artificial Spray Irrigation System

In an attempt to artificially recreate spray to maintain the spray wetlands, NORPLAN began in 1995 to investigate a variety of artificial spray irrigation system options under the Lower Kihansi Long Term Environmental Monitoring Program. The University of Trondheim, Water Construction Faculty, was approached, and several technical solutions were proposed. The options that were considered included (1) a gravity-fed spray irrigation system; (2) a large moveable rotating spray gun; (3) the diversion of water from either the Mhalala stream or from the Kihansi River into a chute or "ski jump" which would generate spray; and (4) fountain jets that pulverized water into spray using a small proportion of the environmental bypass flow in the Kihansi River.

Of these four options, the gravity-fed spray irrigation system and the fountain jets were selected. Diverting water from the Mhalala stream to a "ski jump" proved infeasible because of the discovery of a Kihansi Spray Toad population in the Mhalala wetland located adjacent to the Mhalala stream. A prototype of a large moveable rotating spray gun was tested in the Mid-Gorge Spray Wetland, but this option was abandoned because of mechanical problems and the damage it caused to the vegetation and soil (NORPLAN 2002).

4.3 Intermittent Bypass Flows

An additional non-artificial spray irrigation option that was considered to maintain the spray wetland habitats in the gorge was to conduct short-term intermittent bypass flows from the dam to recreate natural spray across the four wetlands located adjacent to the Kihansi River. In 1999, at the time that NORPLAN proposed this option, little was known as to how large a bypass flow would be required or the duration of a bypass flow to maintain the vegetation and microclimate within the spray wetlands. Furthermore, little if anything was known in 1999 as to how the Kihansi Spray Toad would react to intermittent bypass flow releases. Under the Immediate Rescue and Emergency Measures Project (IREM) that was implemented in 2000 to immediately address the environmental impacts arising from the commissioning of the Lower Kihansi hydropower facility, NORPLAN conducted a series of intermittent bypass flow releases that varied in volume and duration to assess their effects on the spray wetland vegetation, microclimate, and the Kihansi Spray Toad.

4.4 Establishment of a Captive Breeding Population of the Kihansi Spray Toad

In 1999, NORPLAN also proposed as a mitigation measure the establishment of a captive breeding population of the Kihansi Spray Toad. In the late 1990s, NORPLAN contacted the Wildlife Conservation Society about establishing a captive assurance population of the Kihansi Spray Toad at the Bronx and Detroit zoos in the United States, in case the other mitigation measures failed or were insufficient.

NORPLAN contacted the Wildlife Conservation Society because it had an office and staff in Tanzania and because the Wildlife Conservation Society and the Bronx zoo shared the same New York headquarters. In 1999, there was limited expertise and experience worldwide in the captive breeding of tropical amphibians. This, however, was to change rapidly over the following decade due to the emergence of chytridiomycosis that was decimating amphibian species worldwide and particularly in the tropics (Zippel et al. 2011). Since the mid-2000s, there has been an increasing effort by multiple institutions worldwide including zoos, academic institutions, and government bodies to establish captive assurance populations for tropical amphibians. There has also been since the mid-2000s an increasing recognition of the importance of establishing captive assurance populations within the native range of species so as to reduce the risk of accidentally transmitting exotic diseases back into the wild when reintroducing captive populations back to their native range. However, in 2000, no amphibian captive breeding facilities existed in Tanzania nor in Africa. Thus, establishing a captive population of the Kihansi Spray Toad in Tanzania or elsewhere in Africa was not possible.

4.5 Expanded Search for the Kihansi Spray Toad

A further mitigation measure proposed by NORPLAN in 1999 was to assess whether the Kihansi Spray Toad and its spray wetland habitat were present in other locations in the Udzungwa Mountains. If appropriate habitat for the Kihansi Spray Toad could be found in other locations in the Udzungwa Mountains, then NORPLAN proposed to assess the feasibility of translocating animals to these other locations. Under the Lower Kihansi Long Term Environmental Monitoring Program and the subsequent IREM project, a survey was conducted throughout the Udzungwa Mountains for the Kihansi Spray Toad and its critical habitat.

References

Hirji R (2005) Annex 6. Lower Kihansi Environmental Management Project (LKEMP) implementation review mission. Washington, DC

NORPLAN (2002) Lower Kihansi hydropower project immediate rescue and emergency measures. Final report. Report produced for Tanzania Electric Supply Company Ltd (TANESCO), Dar es Salaam, Tanzania

Zippel K, Johnson K, Gagliardo R et al (2011) The amphibian ark: a global community for ex situ conservation of amphibians. Herpetol Conserv Biol 6:340–352

World Bank Environmental Review of the Lower Kihansi Hydropower Project, 2000

5

Abstract

The World Bank initiated in March 2000 a second internal environmental review of the Lower Kihansi hydropower facility. As part of this review, the World Bank conducted three missions to Tanzania in 2000. The objectives of the missions were to meet with stakeholders, assess the status of the Kihansi Spray Toad and its habitat, and review the mitigation measures that NORPLAN proposed to conserve the Kihansi Spray Toad and its habitat. During October 2000, the first population census of the Kihansi Spray Toad in the Kihansi Gorge was conducted. The estimated population of the Kihansi Spray Toad across the five spray wetlands was 12,171 of which 89% of the population was found in the Upper Spray Wetland the largest of the five spray wetlands. Stakeholders at an October 2000 joint donor-Government of Tanzania meeting agreed that establishing a captive assurance population of the Kihansi Spray Toad in US zoos should be the highest priority. It was also agreed that a spray irrigation system should be installed in the Upper, Lower, and Mid-Gorge Spray wetlands and that the current minimum bypass flow of 1.5–1.9 m^3/s be maintained until a final water right could be issued for the operation of the Lower Kihansi hydropower facility.

In response to the concerns raised by Friends of the Earth regarding the likely impact of the Lower Kihansi hydropower facility on the newly described and critically endangered Kihansi Spray Toad, the World Bank initiated in March 2000 a second internal environmental review of the hydropower project. This internal review included three site visits by World Bank teams to Tanzania between April and October 2000. The primary objectives of these visits were to consult with Tanzanian governmental and non-government stakeholders, assess the impact of the

W. D. Newmark, *The Kihansi Spray Toad* (Nectophrynoides asperginis): *Crisis Biology in the 21st Century*, Ecological Studies 254, https://doi.org/10.1007/978-3-031-89599-9_5

commissioning of the dam on the Kihansi Spray Toad and the spray wetland vegetation, and discuss and evaluate NORPLAN's proposed mitigation actions.

In May 2000, I was contacted by the World Bank asking whether I would be willing to participate in the environmental review. I was told by the World Bank that they contacted me because of my long-term field experience in the Eastern Arc Mountains and my background in conservation biology.

5.1 Reduction in Spray Wetland Vegetation

In July 2000, I joined a World Bank team that visited the Kihansi Gorge. This short 2-day visit turned out to be my first of more than 30 visits that I made to the Kihansi Gorge over the following 20 years. During this initial visit, I was immediately struck by the dramatic reduction in the proportion of the spray wetlands that were receiving spray from the minimum bypass flow. Since the commissioning of the hydropower dam in December 1999, the flow in the Kihansi River had declined by 90% (Figs. 5.1 and 5.2).

In a very rapid assessment using a 50 m tape, I estimated a 95% reduction in the spray wetland vegetation receiving spray in the Upper and Lower Spray Wetlands and in the lowest elevational spray wetland (~395 m).

Fig. 5.1 Comparison of the spatial dispersion of spray in the Main Falls under a flow of ~16 m^3/s recorded shown on the left during March 2001 and under a flow of 1.5 m^3/s shown on the right during January 2000. The Main Falls located at an elevation of 1000 m and downstream from the Kihansi hydropower dam is the largest waterfall in the Kihansi Gorge. (Photo credit: TANESCO)

Fig. 5.2 Lower section of the Main Falls under a flow of 19 m^3/s in March 2001. (Photo credit: TANESCO)

The reduction in spray in these three spray wetlands contrasted greatly with what I observed during this same visit to the Mhalala Spray Wetland. The Mhalala Spray Wetland, which is located on a tributary of the Kihansi River, was the only spray wetland in the Kihansi Gorge that had not been impacted by the commissioning of the dam because this spray wetland is maintained by spray generated by a waterfall along the Mhalala stream.

During this July 2000 visit, all of the Kihansi Spray Toad that I observed in the Upper and Lower Spray Wetlands were congregated along the Kihansi River on rock faces kept wet by the minimum bypass flow (Fig. 5.3). In contrast, I observed Kihansi Spray Toads in the Mhalala Spray Wetlands within the spray wetland vegetation as well as on rock faces.

In October 2000 I returned again to Kihansi Gorge as part of a World Bank mission to conduct a more detailed assessment of changes in the spray wetland vegetation resulting from the commissioning of the hydropower plant as well as to estimate the population size of the Kihansi Spray Toad within the Kihansi Gorge.

During the October 2000 field trip to the Kihansi Gorge, I visited all of the spray wetlands in the Kihansi Gorge including a formerly inaccessible rock face at the base of the Main Falls. Peter Hawkes, an entomologist from South Africa, had recently discovered a safe route to access the rock face at the base of the Main Falls.

In the October 2000 survey, I estimated a 93% decrease in the total area of the spray wetlands receiving spray from the bypass flow across the five spray wetlands

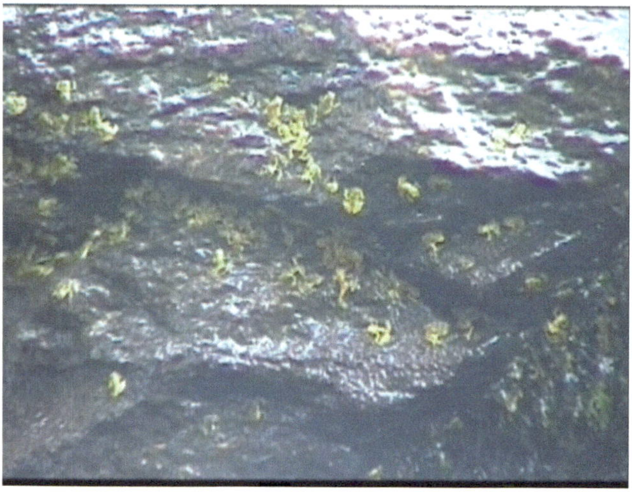

Fig. 5.3 Following the commissioning of the Lower Kihansi hydropower facility and prior to the installation of the spray irrigation systems, nearly all of the Kihansi Spray Toads within the Kihansi Gorge, with the exception of spray toads occurring in the Mhalala Spray Wetland, were congregated on wet rock faces along the Kihansi River. (Photo credit: TANESCO)

in the Kihansi Gorge. These estimates were based on measuring the width and length of each of the spray wetlands and the area of a spray wetland that was receiving spray using a 50 m tape. However, more recent high-resolution aerial imagery of these spray wetlands can now provide a more accurate estimate of the original area of the wetland vegetation within the two largest spray wetlands, the Upper and Lower Spray Wetlands. Based on these more recent high-resolution aerial imagery, an updated estimate of the percent decrease in area of the spray wetlands receiving natural spray in October 2000 was 83% (Table 5.1).

5.2 Kihansi Spray Toad Population, October 2000

In October 2000, I also conducted the first gorge-wide estimate of the population size of Kihansi Spray Toad. Since the discovery of the Kihansi Spray Toad in December 1996, there had not been a population-wide census of the Kihansi Spray Toad in the Kihansi Gorge.

Between 21 and 22 October 2000, I visited all of the spray wetlands having a population of the Kihansi Spray Toad. I conducted a stratified random survey of the Kihansi Spray Toad within the spray wetlands that was stratified by rock face and spray wetland vegetation. On rock faces, toads were sampled in 0.5 m x 0.5 m randomly located plots, and in the spray wetland, vegetation toads were sampled in 0.5 m x 0.5 m plots at 2 m intervals along randomly located line transects. A randomized transect was selected to reduce the impact of human trampling on the spray wetland vegetation. This sampling methodology subsequently became the

Table 5.1 Percent decrease in wetland area receiving natural spray from the minimum bypass flow in October 2000

Spray wetland	Elevation (m)	Original spray wetland vegetation area based on high-resolution aerial imagery (m^2)	Percent of total spray wetland vegetation	Area of spray wetland vegetation receiving natural spray in Oct 2000 (m^2)	Percent decrease in area of wetland receiving natural spray
Main Falls	1030	178	4	178	0
Upper	940	3173	65	578	82
Mhalala	925	218	4	218	0
Lower	850	1114	23	0	100
Mid-Gorge	680	225	5	0	100
Total		**4907**		**840**	**83**

Estimates of the original area of the Upper and Lower Spray Wetlands are based on recent high-resolution Google Earth imagery

standardized sampling protocol for surveying the Kihansi Spray Toad population until the spray toad's population crashed in June 2003. After June 2003, time-constrained survey methods were largely employed because of the general rarity of the Kihansi Spray Toad within all of the spray wetlands.

In October 2000, I estimated the total population of the Kihansi Spray Toad in the Kihansi Gorge was 12, 171 (Table 5.2).

After I completed each of daily surveys in October 2000, Jean-Roger Mercier, the task team leader of the Environmental Review for the World Bank, immediately forwarded the estimates of the Kihansi Spray Toad population to James Wolfenson, the president of the World Bank. This daily reporting to the president of the World Bank highlighted the extreme concern within the World Bank as to the population status and viability of the Kihansi Spray Toad within the gorge.

Nearly 90% of the entire Kihansi Spray Toad population in the Kihansi Gorge in October 2000 was found in the Upper Spray Wetland, the largest of the five spray wetlands (Table 5.2). Yet what was particularly alarming was that more than 80% of the Kihansi Spray Toad population in the gorge was observed during the diurnal surveys on wet rock faces bordering the Kihansi River (Fig. 5.3) that were being kept wet by spray generated from the minimum bypass flow (1.5–1.9 m^3/s). These observations dramatically highlighted the adverse impact that the commissioning of the hydroelectric facility was having on the Kihansi Spray Toad and the importance of establishing as quickly as possible a captive breeding population.

My estimates of the Kihansi Spray Toad density in October 2000 in the Upper Spray Wetland vegetation (4.0 individuals/m^2) were similar to Kim Howell's estimate in the same wetland in October and November 1997 (4.3 individuals/m^2). Kim Howell after discovering the Kihansi Spray Toad in December 1996 returned to the Kihansi Gorge in October and November 1997 to estimate the toad's population in the Upper Spray Wetland. He then extrapolated his estimate of the Kihansi Spray

Table 5.2 Population of the Kihansi Spray Toad in five spray wetlands in the Kihansi Gorge in October 2000

Spray wetland and habitats	Area of habitat (m^2)	Density of toads (no./m^2)	Estimated population (± 95% CI)	Percent of total population
Main Falls				
Rock face	225	2.1	473 ± 441	3.9
Vegetation	42			
Mhalala				
Rock face	28	3.4	95 ± 66	0.8
Vegetation	220			
Upper				
Rock face	284	30.0	8520 ± 4894	70.0
Vegetation	578	4.0	2310 ± 1416	18.9
Lower				
Rock face	110	6.9	759 ± 419	6.2
Vegetation (receiving artificial spray)	400			
Mid-Gorge				
Rock face	59	0.2	12 ± 28	0.1
Vegetation (receiving artificial spray)	210	–	–	
Total			**12,171 ± 5051**	

Source: Newmark (2002)

Toad density in Upper Spray Wetland to the entire gorge. However, my gorge-wide survey in October 2000 suggested that extrapolating the Kihansi Spray Toad density from the Upper Spray Wetland to the entire gorge was quite likely problematic because of the varying density of the Kihansi Spray Toad among the spray wetlands (Table 5.2). Subsequently, Kim Howell informed me that he believed that his gorge-wide estimate of the Kihansi Spray Toad population in 1997 was positively biased and that the October 2000 estimates were quite likely a more accurate estimate of the population size of the Kihansi Spray Toad in the gorge prior to the commissioning of the dam (Kim Howell, *personal communication*).

5.3 Pilot Testing of Spray Irrigation Systems

One of the last activities that I conducted during the October 2000 field trip to the Kihansi Gorge was to inspect the gravity-fed spray irrigation system that had been recently installed and was being tested in the Mid-Gorge Spray Wetland. During my visit to this wetland, all of the Kihansi Spray Toads that I observed were on rock faces that were being kept wet by the artificial spray system. While this was disturbing because it indicated that the minimum bypass flow was insufficient to maintain this spray wetland, it did suggest that rapidly installing an artificial spray irrigation system across the Upper, Lower, and Mid-Gorge Spray Wetlands would be essential to preventing a further reduction in the Kihansi Spray Toad population and spray wetland vegetation in the Kihansi Gorge.

5.4 Review of NORPLAN's Proposed Mitigation Options

During the third week of October 2000, the World Bank mission to Tanzania hosted a stakeholder workshop to review NORPLAN's proposed mitigation measures and to prioritize their implementation. Representatives from the Ministries of Natural Resources and Tourism, Energy, Water, Treasury, the Vice-President's Office, TANESCO, University of Dar es Salaam, NORPLAN, as well as the World Bank and the other co-financing partners attended this workshop.

After considerable discussion, the conclusions of this workshop were: (1) a captive breeding program for the Kihansi Spray Toad be initiated immediately; (2) the current minimum bypass flow of 1.5–1.9. m^3/s be maintained or increased until a final water right for the operation of the dam can be established; (3) the gravity-fed spray irrigation system be rapidly expanded to cover the Mid-Gorge, Lower, and Upper Spray Wetlands; (4) a multi-disciplinary task force be convened to determine the conditions for the final water right for the Lower Kihansi hydropower facility; and (5) an intensified search for the Kihansi Spray Toad be conducted outside of the Kihansi Gorge.

The participants of the workshop agreed that the highest priority should be to establish a captive assurance population of the Kihansi Spray Toad. In October 2000, it was apparent that the Kihansi Spray Toad was facing multiple threats including the permanent drying out of the spray wetlands, a dramatic reduction in food availability, increased risk of fire, and the potential introduction of diseases. It was also highlighted that additional mitigation measures would be required over the medium to long term to conserve the Kihansi Spray Toad and its habitat and that a multi-prong strategy to conserve the Kihansi Spray Toad was essential.

To assist in the evaluation of the proposed mitigation measures, I drafted a qualitative risk assessment that compared the extinction risk of the Kihansi Spray Toad in October 2000 under various combinations of mitigation measures (Table 5.3).

At the end of the workshop, participants selected a medium-risk strategy to reduce the risk of extinction of the Kihansi Toad Spray Toad which involved maintaining a minimum bypass flow of 1.5–1.9 m^3/s, expanding the spray irrigation systems across the largest three spray wetlands, and establishing a captive breeding population.

Although maintaining a minimum environmental bypass flow of 7.7 m^3/s was a stipulation of the provisional water right that the Division of Environment in the Vice-President's Office had issued to TANESCO in November 1996 for the operation of the hydropower plant, because of the daily 4–8-hour rolling electricity blackouts that most of Tanzania was experiencing in October 2000, the government of Tanzania and the donor community agreed that the current bypass flow of

Table 5.3 A qualitative assessment of the extinction risk of the Kihansi Spray Toad in October 2000 under different combinations of NORPLAN's proposed mitigation measures

Extinction risk of the Kihansi Spray Toad								
Very low	Low		Medium			High		Very high
Maintain the historic average discharge of 16.3 m^3/s + captive breeding	Maintain the historic average discharge of 16.3 m^3/s	Maintain a minimum bypass flow of 7.7 m^3/s + spray irrigation + captive breeding	Maintain a minimum bypass flow of 7.7 m^3/s + captive breeding	**Maintain a minimum bypass flow of 1.5–1.9 m^3/s + spray irrigation + captive breeding**	Maintain a minimum bypass flow of 7.7 m^3/s	Maintain a minimum bypass flow of 1.5–1.9 m^3/s + spray irrigation	Maintain a minimum bypass flow of 1.5–1.9 m^3/s + spray irrigation	Maintain a minimum bypass flow of 1.5–1.9 m^3/s

These mitigation measures included varying levels of a minimum bypass flow, combined with or without spray irrigation, and captive breeding. The government of Tanzania and joint-donor community selected a medium-risk mitigation option highlighted in bold to conserve the Kihansi Spray Toad and the spray wetlands

1.5–1.9 m^3/s be in the meantime maintained or increased. A decision regarding the final water right for the operation of the Kihansi Dam would be delegated to a multidisciplinary task force.

5.5 Establishment of a Captive Assurance Population of the Kihansi Spray Toad

Although the establishment of a captive assurance population for the Kihansi Spray Toad was identified in the joint donor-Tanzanian government meeting in October 2000 as the highest priority in case the other proposed mitigation measures to conserve the Kihansi Spray Toad and its habitat were insufficient or to fail, rapidly implementing this mitigation measure proved to be challenging.

At the October 2000 workshop, I strongly supported NORPLAN's recommendation and the government of Tanzania's decision to request assistance from the Wildlife Conservation Society to establish a captive population of the Kihansi Spray Toad at the Bronx and Detroit zoos. The decision to establish a captive breeding population of the Kihansi Spray Toad had been officially endorsed by the government of Tanzania in September 2000 in a letter from the Ministry of Natural Resources and Tourism to the Ministry of Finance (Mercier 2000). Yet, finalizing a Memorandum of Understanding between the Tanzanian government and the Wildlife Conservation Society proved challenging even though the Ministry of Natural Resources and Tourism had stated in February 2000 that negotiations with the Wildlife Conservation Society were in their final stages.

The Tanzania Wildlife Division, which is responsible for providing the necessary CITES export permits to establish a captive breeding population, expressed reservations about exporting a population of the Kihansi Spray Toad to the Bronx and Detroit zoos. The basis for the Wildlife Division's concern was puzzling given that exporting 500 animals from a population of 12,200 would have a minor impact on the Kihansi Spray Toad population in the gorge, particularly in comparison to the impact of the 83% decline in the spray wetland vegetation in the Kihansi Gorge since December 1999. Subsequent discussions revealed that the primary concern of the Wildlife Division related to the ownership of animals that would be exported to the Bronx and Detroit zoos as well as the fair and equitable sharing of any benefits that may potentially arise in the future from genetic resources. These were issues that had been highlighted in the International Convention on Biological Diversity that the Tanzanian government signed and ratified in March 1996.

To resolve this impasse, a high-level meeting was convened on 20 October 2000 following the joint donor-Tanzanian government workshop between representatives of the World Bank which included the Country Director for Tanzania and the Permanent Secretaries in the Vice-President's Office, Ministries of Energy, Water, and Treasury, as well as the Managing Director of TANESCO and representatives from the Ministry of Natural Resources and Tourism. At this meeting, the Country Director of the World Bank confirmed his availability to facilitate negotiations between the Government of Tanzania and the Wildlife Conservation Society which highlighted the importance from the perspective of the World Bank of initiating a captive breeding program for the Kihansi Spray Toad as quickly as possible.

Three days after this meeting, Tanzanian governmental authorities invited the Wildlife Conservation Society to come to Dar es Salaam to resolve the remaining outstanding issues and to sign a legal agreement that would permit the initiation of a captive breeding program for the Kihansi Spray Toad in the US zoos. An important condition of this agreement was that the Government of Tanzania could use the progeny from the captive breeding program in the United States to support future in situ conservation efforts for the Kihansi Spray Toad in Tanzania.

In late November 2000, representatives from the Wildlife Conservation Society collected 499 adult Kihansi Spray Toads in the Kihansi Gorge. On 30 November 2000 the Bronx zoo received 269 adults, and the National Amphibian Conservation Center at the Detroit Zoo received 230 adults (Searle 2001), thus initiating the captive assurance population of the Kihansi Spray Toad in the US zoos.

5.6 Mobilizing Financial Resources

In October 2000, the Government of Tanzania formally requested support from the donor community to address the environmental impacts associated with the Lower Kihansi Hydropower Project. In the Tanzanian government's request, it was noted that both short- and medium-term support would be required to address the full range of environmental impacts arising from the Lower Kihansi Hydropower Project.

In November 2000, the World Bank issued a summary of its internal environmental review of the Lower Kihansi Hydropower Project (World Bank 2000). This summary listed the mitigation measures that were jointly agreed upon by the October 2000 workshop participants and highlighted the need to mobilize financial resources over the short and medium term to implement critical mitigation measures to conserve the Kihansi Spray Toad and its habitat.

In December 2000, NORAD and SIDA agreed to co-fund a 1-year Immediate Rescue Emergency Measures Project (IREM). The IREM project was designed as a bridging project to allow time for the World Bank to develop a medium-term project that could address more broadly the environmental impacts that were associated with the Lower Kihansi Hydropower Project.

The objective of IREM was to implement short-term mitigation measures essential to conserving the Kihansi Spray Toad and its habitat. The $3.1 million IREM project formally began in March 2001. However, due to the urgency of implementing the identified short-term mitigation measures, NORPLAN initiated IREM in late 2000.

References

Mercier JR (2000) Lower Kihansi Hydropower Project (LKHP)—Tanzania Power VI: Chronology of biodiversity, water rights and mitigative measures. Internal non-official World Bank report

Newmark WD (2002) Conserving biodiversity in East African forests: a study of the Eastern Arc Mountains. Springer, Berlin

Searle J (2001) Kihansi Spray Toad report #2. Wildlife Conservation Society, Department of Herpetology, Bronx, New York

World Bank(2000) Draft final aide memoire. Lower Kihansi Hydropower Project Environmental review. October 2000 Mission. Washington, D.C.

Immediate Rescue and Emergency Measures Project, 2000–2001

6

Abstract

Under the 1-year Immediate Rescue Emergency Measures Project (IREM), six activities were implemented to conserve the Kihansi Spray Toad and its habitat: (1) an expansion of the pilot spray irrigation systems to the Upper, Lower, and Mid-Gorge Spray Wetlands; (2) construction of walkways, bridges, and ladders in the Kihansi Gorge; (3) an assessment of the effectiveness of short-term intermittent flow releases in maintaining the spray wetlands; (4) ecological studies of the Kihansi Spray Toad and its food and habitat; (5) an expanded search for potential translocation sites for the Kihansi Spray Toad outside of the Kihansi Gorge; and (6) development of a Kihansi area conservation plan.

In November 2000, NORPLAN at the request of TANESCO presented a proposal to the joint donors to implement a series of immediate mitigation measures to conserve the Kihansi Spray Toad and spray wetlands. These measures included (1) an expansion of the pilot spray irrigation systems to the Upper, Lower, and Mid-Gorge Spray Wetlands; (2) construction of walkways, bridges, and ladders in the gorge to enhance worker safety and to minimize human trampling in the wetlands; (3) an assessment of the effectiveness of short-term intermittent flow releases in maintaining the spray wetlands; (4) ecological studies of the Kihansi Spray Toad and its food and habitat; (5) an expanded search for potential translocation sites for the Kihansi Spray Toad outside of the Kihansi Gorge; and (6) development of a Kihansi Area Conservation Plan. In addition, NORPLAN conducted at the request of the Tanzanian government a capacity needs assessment and the initial drafting of terms for the final water right as input to the planned World Bank medium-term Lower Kihansi Environmental Management Project (LKEMP).

W. D. Newmark, *The Kihansi Spray Toad* (Nectophrynoides asperginis): *Crisis Biology in the 21st Century*, Ecological Studies 254, https://doi.org/10.1007/978-3-031-89599-9_6

6.1 Expansion of the Pilot Spray Irrigation Systems

In June 2000, NORPLAN initiated the testing of a gravity-fed spray irrigation system in the Kihansi Gorge. The spray irrigation system consisted of multiple parallel lines of plastic piping and nozzles mounted 1–2 m above the ground. The plastic piping lines were spaced approximately 5 m apart across a spray wetland, and nozzles were placed at 3.5 m intervals along the plastic piping (Fig. 6.1). The sprinkler nozzles were designed to mimic pre-diversion natural spray in terms of droplet size and density.

The first tests of the pilot spray irrigation system were conducted in the Mid-Gorge Spray Wetland comprising of three sprinkler lines which covered an area of 350 m^2 or 40% of the wetland (NORPLAN 2001). Fifteen months later in September 2001, NORPLAN expanded the size of this pilot gravity-fed sprinkler irrigation system in the Mid-Gorge Spray Wetland to five sprinkler lines which covered an area 420 m^2.

The Lower Spray Wetland was the second spray wetland in which the gravity-fed spray irrigation system was installed. In September 2000, NORPLAN installed 15 sprinkler lines that covered 1400 m^2 in the Lower Spray Wetland (Fig. 6.1). The Upper Spray Wetland was the last of the three large spray wetlands in which the gravity-fed spray irrigation system was installed. In March 2001, NORPLAN installed 17 sprinkler lines in the Upper Spray Wetland covering 2900 m^2.

NORPLAN also installed two fountain jet spray irrigation systems adjacent to the Upper Spray Wetland. The first and larger of the two fountain jet spray irrigation systems was installed at the base of the Main Falls, and the second and smaller fountain jet spray irrigation system was installed 25 m above and to the west of the base of the Main Falls (Fig. 6.2). The spray that was generated by the fountain jet spray irrigation systems was designed to be dispersed over the Upper Spray Wetland by wind that funneled through the gorge.

The spray irrigation systems that were installed in the Upper, Lower, and Mid-Gorge Spray Wetlands were designed to mimic as closely as possible the droplet size and volume of water that had been generated prior to the commissioning of the

Fig. 6.1 The gravity-fed sprinkler system in the Lower Spray Wetland

Fig. 6.2 The small and large fountain jets, shown on the left and right, respectively, were installed along the base of the Main Falls to provide additional spray for the Upper Spray Wetland. (Photo credit: TANESCO)

dam by waterfalls located adjacent to these three wetlands. Earlier analyses by NORPLAN and their consultants found that under a natural flow of 20 m^3/s in the Kihansi River, the portions of the Upper Spray Wetland located close to the base of the Main Falls received a minimum daily volume of 1145 mm (NORPLAN 2002a) while the center of the Upper Spray Wetland received a volume of approximately 76 mm of water (Taplin et al. 1999). NORPLAN's gravity-fed spray irrigation system was designed to deliver a minimum volume 70 mm of water per day.

One of the initial challenges of the gravity-fed spray irrigation system was the clogging of the nozzles. This problem was addressed by installing sediment traps and in-line filters. Nonetheless, the gravity-fed spray irrigation systems in the three spray wetland required regular servicing. Under the medium-term follow-on project to IREM (the Lower Kihansi Environmental Management Project), TANESCO hired a team of technicians to weekly service and repair the spray irrigation systems in the Kihansi Gorge. Having an abundant supply of parts proved to be essential to the operation of the spray irrigation system because of the vulnerability of the gravity-fed spray irrigation system to floods, landslides, and tree falls, all of which damaged portions of the spray irrigation systems in the spray wetlands over the following 20 years.

6.2 Short-Term Effects of the Spray Irrigation Systems on the Spray Wetlands

Due to the importance of pre-testing the gravity-fed spray irrigation system in the Mid-Gorge Spray Wetland before deploying it across the Lower and Upper Spray Wetlands, there was a 9-month delay following the commissioning of the dam before the gravity-fed spray irrigation system was deployed in the Lower Spray Wetland and a 14-month delay in the Upper Spray Wetland. The longer delay in the deployment of the spray irrigation system in the Upper Spray Wetland has been suggested as a possible contributary factor to the longer recovery time in the plant community of the Upper Spray Wetland following the commissioning of the Lower Kihansi hydropower facility than was observed in the Lower Spray Wetland.

On the other hand, as will be discussed later, the plant community structure in the Mid-Gorge Spray Wetland, where the spray irrigation system was first installed in the Kihansi Gorge, has recovered less after 20 years than the plant community structure in the Upper and Lower Spray Wetlands (Sect. 13.1). However, soil erosion in the Mid-Gorge Spray Wetland resulting from the initial testing of the spray irrigation systems in this wetland has very likely been a contributory factor in the slower recovery of the plant community structure in this spray wetland. In 1999, NORPLAN tested a large moveable spray gun irrigation system in the Mid-Gorge Spray wetland. This testing, however, was discontinued because the large moveable spray gun irrigation system caused extensive soil erosion and damaged the spray wetland vegetation.

6.3 Construction of Walkways, Bridges, and Ladders

To reduce the impact of human trampling of the spray wetland vegetation and to enhance wetland accessibility as well as the safety of maintenance staff servicing the spray irrigation infrastructure, NORPLAN constructed under the IREM project a system of walkways, bridges, ladders, and staircases throughout the gorge.

The spray wetland vegetation and soils, particularly in the wetter portions of the spray wetlands, were historically very spongy, and when fully hydrated, a person walking through the spray wetlands could easily sink up to their ankle. Repeated walking along the same path in the spray wetlands would extensively damage if not destroy the vegetation (Fig. 6.3). To reduce human impacts on the spray wetland vegetation, NORPLAN constructed walkways using concrete steps in the Upper, Lower, and Mid-Gorge Spray Wetlands (Fig. 6.4). The walkways were positioned to allow researchers to access monitoring sites and to permit the maintenance staff to more easily service the spray irrigation systems.

NORPLAN also constructed under the IREM project three suspension bridges (Fig. 6.5) across the Kihansi River and a series of wooden ladders (Fig. 6.6) and staircases (Fig. 6.7) that were placed along wet and exposed rock faces within the gorge so to enhance spray wetland accessibility and maintenance staff safety in servicing the spray irrigation systems.

Fig. 6.3 Human trampling in the Upper Spray Wetland prior to the installation of walkways. (Photo credit: TANESCO)

Fig. 6.4 A concrete block walkway leading to the entrance of the Lower Spray Wetland

6.4 Intermittent Bypass Flow Studies

NORPLAN also initiated under IREM an assessment of the value of short-term intermittent bypass flow releases from the dam as a mitigation measure to conserve the spray wetland vegetation in case the pilot spray irrigation systems proved ineffective or were temporally disabled. Between 2000 and 2001, NORPLAN assessed whether intermittent discharges of 4 m^3/s and 8 m^3/s over a period of 2–4 hours per

Fig. 6.5 One of three suspension foot bridges that spans the Kihansi River. This bridge provides accesses to the Mhalala Spray Wetland

Fig. 6.6 An access ladder constructed to service the water supply lines for the Upper Spray Wetland sprinkler systems. (Photo credit: TANESCO)

Fig. 6.7 Access stairway to the Jibali intake for the Upper Spray Wetland irrigation system

day and over 5–9 days would provide adequate spray to maintain the microclimate within spray wetlands and how the Kihansi Spray Toad would respond to these discharges.

Results from these initial assessments that were conducted in the Upper Spray Wetland indicated that the spray toads remained on wet rocks faces under a discharge of 4 m^3/s but moved from wet rock faces along the river into the spray wetland vegetation under a discharge of 8 m^3/s. However, it was not possible to assess how the spatial dispersion of the spray varied across the spray wetlands under these two experimental discharges.

As a result, NORPLAN proposed to conduct additional intermittent flow studies to assess spray dispersion across a wider range of discharges. Between 2002 and 2003, NORPLAN examined effects of discharges of 4 m^3/s, 6 m^3/s, and 8 m^3/s for 3–4 hours per day over 3–5 days on spray dispersion. NORPLAN also examined during this time period the effects of discharges of 12 m^3/s and 16 m^3/s over 4 hours during a single day on spray dispersion.

These experimental discharges revealed that only discharges of 12 m^3/s and 16 m^3/s produced sufficient spray to entirely cover the Upper Spray Wetland. However, interpreting the impact of the intermittent discharges on temperature, relative humidity, and soil moisture proved difficult because of the many uncontrolled and confounding factors the influence spray dispersion, temperature, and relative humidity. Additionally, TANESCO was reluctant to release large volumes of water from the dam for these larger intermittent discharge experiments because of the ongoing power shortages that were occurring throughout Tanzania during 2002 and 2003. Consequently because of the difficulty in conducting and interpreting these quasi-experiments and TANESCO's reluctance to release large volumes of water, it was concluded that intermittent discharges were not a viable mitigation option for conserving in situ the Kihansi Spray Toad and the spray wetland habitat.

6.5 Expanded Search for the Kihansi Spray Toad

NORPLAN had conducted between 1998 and 1999 under the auspices of the Lower Kihansi Long Term Environmental Monitoring Program a search for the Kihansi Spray Toad within 23 km of the Kihansi Gorge. At the request of the Tanzanian government, NORPLAN conducted under the IREM project an expanded search for the Kihansi Spray Toad and its habitat within 50 km of the Kihansi Gorge. A secondary objective of this search was to identify potential translocation sites for a Kihansi Spray Toad population, if suitable habitat could be found elsewhere in the Udzungwa Mountains.

The search conducted under IREM was focused on the eastern escarpment of the Udzungwa Mountains and combined aerial and ground surveys. The aerial surveys were conducted during the height of the dry season to ensure that the previously unvisited or unknown waterfalls were perennial rather than intermittent. A total of 13 waterfalls were located of which 6 of these waterfalls had not been visited during the 1998–1999 surveys (Table 6.1).

Between May 2000 and November 2001, field biologists visited all 13 waterfalls that were located within 50 km of the Kihansi Gorge. All of the waterfalls were visited on one or more occasions during the day and several at night. During a visit, an inventory was conducted of all anuran species residing in close proximity to the falls as well as the presence/absence of spray wetland vegetation and the daily spray rate for the largest waterfalls. Additionally, at the Upper and Lower Lufulutonya Spray Wetlands, which are located 2 km upstream from the Kihansi hydroelectric dam, a survey of insect fauna and hourly recordings of temperature and relative humidity were conducted over a 3-week period.

None of the field surveys conducted between 1998 and 2001 recorded the presence of the Kihansi Spray Toad at any of 13 waterfalls, although 5 other anuran species were documented across these 13 sites (Table 6.1). Few of the visited sites contained spray wetland vegetation, and for those that did, the area of the spray wetland vegetation was small (Table 6.1). Thus, NORPLAN concluded that translocating the Kihansi Spray Toad to a new site outside of the gorge was infeasible, and consequently this mitigation option was excluded from further consideration under the Lower Kihansi Environmental Management Project.

Table 6.1 Ecological characteristics of 13 sites in the Udzungwa Mountains located within 50 km of the Kihansi Gorge containing a waterfall (NORPLAN 2002a)

Location	Drainage	Other anuran species	Presence of spray wetland vegetation	Estimated spray rate (mm/day)	Survey dates
Upper Lufulutonya	Kihansi River	*Arthroleptis* sp. *Afrana angolensis* *Arthroleptides* sp.	Yes	≥300	Mar 1998 Mar 1999 Jun 2000 May 2001
Lower Lufulutonya	Kihansi River	*Afrana angolensis* *Arthroleptides* sp.	Yes	≥300;	Mar 1998 Mar 1999 Jun 2000 May 2001
Udagaji Gorge	Udagaji River	*Nectophrynoides tornieri* *Arthroleptides* sp. *Leptopelis uluguruensis*	Yes	> 58 mm;	Jun 1998 Jun 2000 Jun 2001
Masisiwe Gorge—Lower	Mkaja River	*Arthroleptides* sp. *Arthroleptis stenodactylus* *Schoutedenella xenodactyla*	No		Jul 1998 May 2001
Sanje Falls—Upper	Sanje River	*Arthroleptides* sp. *Leptopelis uluguruensis*	Yes		1998[a] Oct 2001
Sanje Falls—Middle	Sanje River	*Arthroleptides* sp.	No		1998[a] Oct 2001
Sanje Falls—Lower	Sanje River	*Arthroleptides* sp.	Yes		1998[a] Oct 2001
Kiselesi	Kiselesi River	*Arthroleptides* sp. *Schoutedenella xenodactyla*	No	6	Nov 2001
Igolimbogo	Kiselesi River	*Arthroleptides* sp.	No		Nov 2001
Kifufu	Kiselesi River	*Arthroleptides* sp.	No		Nov 2001

(continued)

Table 6.1 (continued)

Location	Drainage	Other anuran species	Presence of spray wetland vegetation	Estimated spray rate (mm/day)	Survey dates
Izanga	Izanga River	None	No		Nov 2001
Msingusi	Msingusi River	None	No		Nov 2001
Ungwilo	Ungwilo River	None	No		Nov 2001

These sites were surveyed between 1998 and 2001 for the presence of the Kihansi Spray Toad and as potential translocation sites
[a]Month was not recorded in the summary report (NORPLAN 2002a)

6.6 Ecological Studies

6.6.1 Kihansi Spray Toad and Other Amphibian Species

One of the more important project activities implemented under the IREM project was to conduct an expanded ecological study of the Kihansi Spray Toad that had been initiated under the Lower Kihansi Long-term Environmental Monitoring Program. Alan Channing a herpetologist from the University of the Western Cape in South Africa was commissioned to conduct this study. His resulting study (Channing et al. 2006) forms the basis for much of what we know about the natural history and ecology of the Kihansi Spray Toad in the wild (Sects. 3.1, 3.2, 3.3, 3.4, 3.5, 3.6, 3.7 and 3.8). His study also provided estimates between February 2001 and March 2002 of the population of the Kihansi Spray Toad within the Kihansi Gorge (Sect. 8.3).

6.6.2 Entomological Studies

Another important concern arising from the commissioning of the Lower Kihansi hydropower facility in December 1999 was the impact of a 90% reduction in the flow of the Kihansi River through the gorge on the invertebrate fauna within the spray wetlands. Under the Lower Kihansi Long Term Environmental Monitoring Program, Peter Hawkes, an entomologist from South Africa, had initiated a study to identify the diet of the Kihansi Spray Toad and to assess the abundance of important groups of insects in the spray wetlands. Peter Hawkes continued with this assessment of the invertebrate fauna in the spray wetlands under the IREM project.

An important result that emerged under the IREM project based on an inspection of stomach contents of the Kihansi Spray Toad was that the spray toad is a generalist feeder (NORPLAN 2002b). Peter Hawkes identified 18 orders of arthropods in stomach contents of which 88.5% were insects. The largest group of insects were Diptera, Homoptera, and Hymenoptera.

Under the IREM project, Peter Hawkes selected two invertebrates, *Afrosteles distans* in the family Cicadellidae and a undescribed species of *Ortheziola* in the family Ortheziidae, to monitor as indicator species of food abundance in the spray wetlands. These two invertebrates were selected because they were the most commonly identified species in the stomach content analysis of the Kihansi Spray Toad, comprising 9.4% of the spray toad's diet by number. These two invertebrates were also extremely abundant in the spray wetland vegetation prior to the commissioning of the dam (NORPLAN 2002b).

The two indicator invertebrates species differed in terms of movement and habitat preference. *Afrosteles distans* is fast-moving and both jumps and flies and is most commonly found on erect vegetation within the spray wetlands. In contrast, *Ortheziola* sp. is slow moving and is predominantly found along the base and exposed roots of plants within the spray wetlands.

Peter Hawkes developed a novel method to monitor the abundance of *Afrosteles distans*. This entailed dropping a metal frame over wetland vegetation at random locations and recording the number of individuals captured within a dish containing ethanol. *Ortheziola* sp., which is slow moving, was surveyed by recording the number of individuals observed and collected over a 5-minute period within a 15 cm x 15 cm quadrat.

Comparison of the abundance of these two indicator species in the Upper Spray Wetland pre- and post-diversion revealed that the installed spray irrigation systems were maintaining populations of both species (Sect. 13.3). However, *Afrosteles distans* responded more quickly and positively to the installation of the spray irrigation systems than *Ortheziola* sp. (NORPLAN 2002b).

6.6.3 Spray Wetland Vegetation

An additional concern about the sudden reduction in flow of the Kihansi River through the gorge was its impact on the spray wetland vegetation, the critical habitat of the Kihansi Spray Toad. Under IREM, a study was initiated that compared the vegetation community structure in the Upper Spray Wetland prior to the reduction in flow of the Kihansi River to that which occurred following the installation of the spray irrigation system in June 2001. In addition, plots in the Upper Spray Wetland that received artificial spray were compared with plots in the Upper Spray Wetland that did not receive artificial spray (NORPLAN 2002a).

The drying out of the spray wetland between the commissioning of the dam in December 1999 and the installation of the spray irrigation system in the Upper Spray Wetland in June 2001 resulted in the disappearance of three plant species and an increase in abundance of grass species. However, these changes following the commissioning of the Kihansi in the vegetation community did not appear to have impacted food abundance for the Kihansi Spray Toad because one of the grass species that increased in abundance *Microstegium vagans* is a dominant food plant of *Afrosteles distans*, one of the two indicator invertebrate species in the diet of the Kihansi Spray Toad.

Under the follow-on projects to IREM, the monitoring of the Kihansi Spray Toad and its indicator food items, as well as the vegetation composition and structure of the spray wetlands, continued.

6.7 Kihansi Area Catchment Plan

Finally, under the IREM project support was provided for the development of a catchment management plan for the Kihansi River. The objectives of this plan were to propose solutions to (1) resolve impacts arising from changes in land use in the area immediately surrounding the gorge; (2) maintain hydrologic flows to the hydropower facility; (3) maximize benefits to local populations and encourage their participation in the management of the area; and (4) ensure the long-term sustainability of the Kihansi catchment and natural resources. The resulting Kihansi Area Catchment Plan that was developed under IREM was the first of three plans that were developed over 20 years to manage the Kihansi Gorge and surrounding area.

References

Channing A, Finlow-Bates KS, Haarklau SE, Hawkes PG (2006) The biology and recent history of the critically endangered Kihansi Spray Toad Nectophrynoides asperginis in Tanzania. J East African Nat Hist 95:117–138. https://doi.org/10.2982/0012-8317(2006)95[117:tbarho]2.0.co;2

NORPLAN (2001) Lower Kihansi hydropower project. Immediate rescue and emergency measures. Mid-term report. Report produced for Tanzania Electric Supply Company Ltd (TANESCO), Dar es Salaam, Tanzania

NORPLAN (2002a) Lower Kihansi hydropower project immediate rescue and emergency measures. Final report. Report produced for Tanzania Electric Supply Company Ltd (TANESCO), Dar es Salaam, Tanzania

NORPLAN (2002b) Lower Kihansi hydropower project. Immediate rescue and emergency measures. Final specialist report: entomological studies. Report produced for Tanzania Electric Supply Company Ltd. (TANESCO), Dar es Salaam, Tanzania

Taplin J, Lovett JC, McLean C (1999) Conservation status and moisture requirements of the vegetation of the Kihansi Gorge vegetation. Report prepared for NORPLAN/TANESCO, Dar es Salaam, Tanzania

Design of the Lower Kihansi Environmental Management Project, 2001

7

Abstract

The Lower Kihansi Environmental Management Project was designed in 2001 to address the medium-term environmental impacts arising from the Lower Kihansi Hydropower Project. Given the multiplicity of threats that were facing the Kihansi Spray Toad and its critical habitat in the early 2000s, developing and implementing a Kihansi Spray Toad recovery program would require the involvement of multiple governmental and non-governmental stakeholders. To reconcile competing mandates and interests among stakeholders, a high-level governmental multi-agency Steering Committee was established chaired by the Permanent Secretary of the Vice President's Office. To assist the Steering Committee in decision-making, a Multi-sectoral Technical Advisory Committee was established consisting of representatives from key ministries. Activities that were identified to be implemented during the first year of LKEMP included initiating a review of the previous research and ecological monitoring in the Kihansi Gorge, identifying future research and monitoring priorities, and developing a training program to enhance national capacity in conservation biology, water resource and environmental management, and captive breeding of amphibians and amphibian husbandry.

In October 2000, the government of Tanzania requested support from the joint-donor community to address the environmental impacts associated with the Lower Kihansi Hydropower Project. The World Bank agreed to develop a medium-term, 5-year project to address the environmental impacts arising from this project. In order to give the World Bank sufficient time to this develop a project, the Swedish and Norwegian development agencies agreed to fund a 1-year Immediate Rescue

W. D. Newmark, *The Kihansi Spray Toad* (Nectophrynoides asperginis): *Crisis Biology in the 21st Century*, Ecological Studies 254, https://doi.org/10.1007/978-3-031-89599-9_7

Emergency Measures Project (IREM) to conserve the Kihansi Spray Toad and its habitat (Chap. 6).

In January 2001, the World Bank initiated the development of the Lower Kihansi Environmental Management Project (LKEMP), a 5-year project focused on addressing the medium-term environmental impacts arising from the Lower Kihansi Hydropower Project. LKEMP was designed to not only extend the mitigation activities initiated under the IREM project to conserve the Kihansi Spray Toad and its critical habitat but also to address more broadly the environmental impacts arising from the Lower Kihansi Hydropower Project.

An important consideration in the design of the LKEMP and IREM was the temporal sequencing of mitigation measures to conserve the Kihansi Spray Toad and its critical habitat. To help guide project designers and implementers, I developed in July 2000 a guideline for the temporal implementation of short-, medium-, and long-term mitigation measures to conserve biodiversity within the Kihansi gorge (Table 7.1).

Included in this guideline was the identification of potential risks and critical information required to assess the effectiveness of the proposed mitigation measures. Between 2000 and 2012, much of IREM and LKEMP's project activities were focused on assessing the short- and medium-term effectiveness of the conservation mitigation measures that had been proposed under the Lower Kihansi Hydropower Project.

As outlined in Table 7.1, the highest priorities over the short term were the establishment of a captive breeding population and the expansion of the spray irrigation system across the Upper, Lower, and Mid-Gorge Spray Wetlands. This would be followed over the medium term by an expansion of the walkways within the spray wetlands, an expanded assessment of the value of intermittent flows, a survey of the distribution and the abundance of several newly discovered plant species, and an assessment of the presence/absence of other potential translocation sites in the Udzungwa Mountains. Lastly it was proposed that over the medium to long term, a landscape-wide conservation plan for the Kihansi Gorge should be developed as part of a larger conservation strategy for the Udzungwa Mountains.

7.1 Supporting Elements of the Species Recovery Program

7.1.1 Stakeholder Involvement

Given the multiplicity of threats that were facing the Kihansi Spray Toad and its habitat in 2000, it was apparent that developing and implementing a Kihansi Spray Toad recovery program would require the involvement of multiple governmental and non-governmental stakeholders. The most immediate threats in 2000 to the Kihansi Spray Toad and its habitat arose from the competing demands for water in the Kihansi catchment, specifically, water to maintain critical habitat for the Kihansi Spray Toad versus that required to operate the recently completed Lower Kihansi hydropower facility.

Table 7.1 Proposed temporal sequence for implementation of conservation mitigation measures in the Kihansi Gorge and their associated potential risks and critical missing information

Conservation mitigation measure	Current status	Expected benefits	Beneficial timeframe	Potential risks	Critical missing information
1. Establish a captive breeding population of the Kihansi Spray Toad	Ongoing	"Insurance" policy" against the extinction of the toad	Short to medium term	Limited capacity to maintain captive-bred species in perpetuity	Survivorship of toads in captivity
2. Installation of an artificial spray irrigation system (a) Spray misting (b) Fountain jet	(a) Being tested (b) Preliminary feasibility assessment	Maintain spray wetland during the dry season	Short to medium term	Disturbance of habitat during installation; long-term maintenance	Response of the Kihansi Spray Toad, spray wetland vegetation, insects, and predators
3. Establish permanent walkways within the spray wetlands	Preliminary assessment	Reduce human disturbance in the spray wetlands	Short to medium term	May be placed incorrectly for purposes of monitoring; damage wetland during installation	Indirect effects on the Kihansi Spray Toad and wetland vegetation
4. Intermittent bypass flows	Limited observations	Maintain spray wetland and montane forest during the dry season	Short to medium term	Unwillingness or inability of TANESCO to release desired bypass flows	Response of spray toads and their food items to rapid fluctuations in discharge and their effectiveness in reducing habitat desiccation
5. Survey distribution and abundance of newly discovered plant species	Limited observations	Establish the level of threat facing the new plant species	Short to medium term	May overlook sites containing these species	Many species have not been formally described
6. Survey distribution and abundance of the Kihansi Spray Toad and appropriateness of habitat outside the Kihansi Gorge	Limited observations	Identification of other populations and sites for translocation	Short to medium term	May overlook appropriate sites	The likelihood of the of Kihansi Spray Toad having a highly disjunct distribution

(continued)

Table 7.1 (continued)

Conservation mitigation measure	Current status	Expected benefits	Beneficial timeframe	Potential risks	Critical missing information
7. Assess the feasibility of translocating the Kihansi Spray Toad to an additional site	No action to date	Reduce the risk of extinction of the toad	Short to medium term	Translocations of amphibians have had a high failure rate	Detailed ecology of the toad and identification of suitable sites
8. Landscape-wide conservation plan for the Kihansi Gorge in the context of the Udzungwa Mountains that would include expanding protected area coverage	No action to date	Reduce long-term risk of extinction for many plant and animal species in the Kihansi Gorge	Medium to long term	Unknown social and economic costs associated with expanding protected areas	Social and economic costs of expanding protected area coverage

This assessment was developed in February 2001 for purposes of assisting the design and implementation of the proposed mitigation measures to conserve plant and animal species within the Kihansi Gorge

These competing demands were exacerbated by the rolling electricity blackouts that much of the country was experiencing in 2000. In the early 2000s, less than 10% of Tanzania's population had access to electricity. The newly commissioned Lower Kihansi hydropower facility was not only the most efficient hydropower facility in Tanzania in terms of electricity generation per cubic meter of water but in 2000 was providing 31% of all of Tanzania's electrical generation. Between 2000 and 2005, the Lower Kihansi hydropower facility generated 23% of Tanzania's total electricity (Hirji 2006).

In designing a species recovery program for the Kihansi Spray Toad, it was essential to involve not only governmental and non-governmental stakeholders having responsibilities and interests for managing wildlife and natural resources in Tanzania but also governmental and non-government stakeholders concerned with water, energy, and agriculture.

To reconcile competing mandates and interests among the many stakeholders, a high-level governmental multi-agency Steering Committee was established comprising of the Permanent Secretaries of the Vice President's Office (Chairperson) and the Ministries of Natural Resources and Tourism, Energy and Minerals, Water and Livestock Development, and Finance. The responsibility of the Steering Committee was to provide overall supervision and oversight of LKEMP and resolve competing interests and tradeoffs among stakeholders in the Kihansi catchment.

To assist the Steering Committee in decision-making, a Multi-sectoral Technical Advisory Committee (MTAC) was established. This committee consisted of representatives from the Division of Environment (Vice President's Office), National Environmental Management Council, the Ministry of Natural Resources and Tourism, TANESCO, the Ministries of Water and Livestock Development, and the Ministry of Energy and Minerals.

Additionally, local and regional governmental authorities as well as national and international academic and non-governmental conservation organizations were also important stakeholders and were actively involved in the design and implementation of key elements of the Kihansi Spray Toad recovery program.

The National Environmental Management Council (NEMC) was tasked with leading project implementation because its legal mandate under Tanzanian law was to coordinate environmental policies and programs in Tanzania. A project coordinator, who was housed in NEMC, was employed by the project to lead the day-to-day implementation of the project.

Central to implementing a Kihansi Spray Toad recovery program was establishing close coordination and support among the key institutions and individuals responsible for the day-to-day implementation of LKEMP. These included the project coordinator and director general of NEMC on the Tanzanian side and LKEMP's task leader and team members on the World Bank side.

The capacity and effectiveness of implementing key project activities in a coordinated and timely fashion were in turn highly dependent upon the experience and expertise of these key institutions and individuals. Over the two decades of implementation of the Kihansi Spray Toad recovery program, all but one individual in these key positions changed over time. These changes in key personnel did influence the culture of the project but were not in my opinion an important determinant in the achievement of the overarching objective of the Kihansi Spray Toad recovery program—the establishment of a self-sustaining population of the Kihansi Spray Toad in the Kihansi Gorge.

7.1.2 Institutional Inputs to the Species Recovery Program

More than 30 national and international institutions provided major inputs between 2001 and 2019 in designing and implementing major components of the Kihansi Spray Toad recovery program (Fig. 7.1). During the first decade of implementation of the Kihansi Spray Toad recovery program, international institutions and consultants with expertise in captive breeding reintroduction of amphibian species and ecological monitoring played a key role in designing project activities and training national scientists. During the second decade of implementation, national scientists were almost entirely responsible for the design and implementation of the species recovery program.

7.2 Research and Monitoring

Research and monitoring were central to the design and implementation of a species recovery program for the Kihansi Spray Toad in large part because there is no "cook book" for designing and implementing species recovery programs. A collaborative, experimentally driven, and adaptive management approach—which involved a continuous review and refinement of research and monitoring based on prior experience—was adopted. Essential to conducting an adaptive management approach was the free and regular exchange of information and data among institutions, scientists, and project personnel involved in implementing the species recovery program.

Yet, given the varying backgrounds, experience, and training among the many institutions and individuals involved in designing and implementing the species recovery program for the Kihansi Spray Toad, there were inevitable differences in opinion about the design and implementation of the species recovery program.

The design and implementation of an ecological monitoring program for the Kihansi Spray Toad and its critical habitat were particularly challenging because of the many institutions that were responsible for implementing the ecological monitoring program. Extensive discussions conducted between 2003 and 2006 were focused on the identification of key ecological indicators, and the frequency and methods with which these indicators would be monitored (Sect. 8.2).

7.3 Training

Central to implementing a species recovery program for the Kihansi Spray Toad and to enhance its long-term sustainability was the development of a training program focused on students, national scientists, and governmental officials.

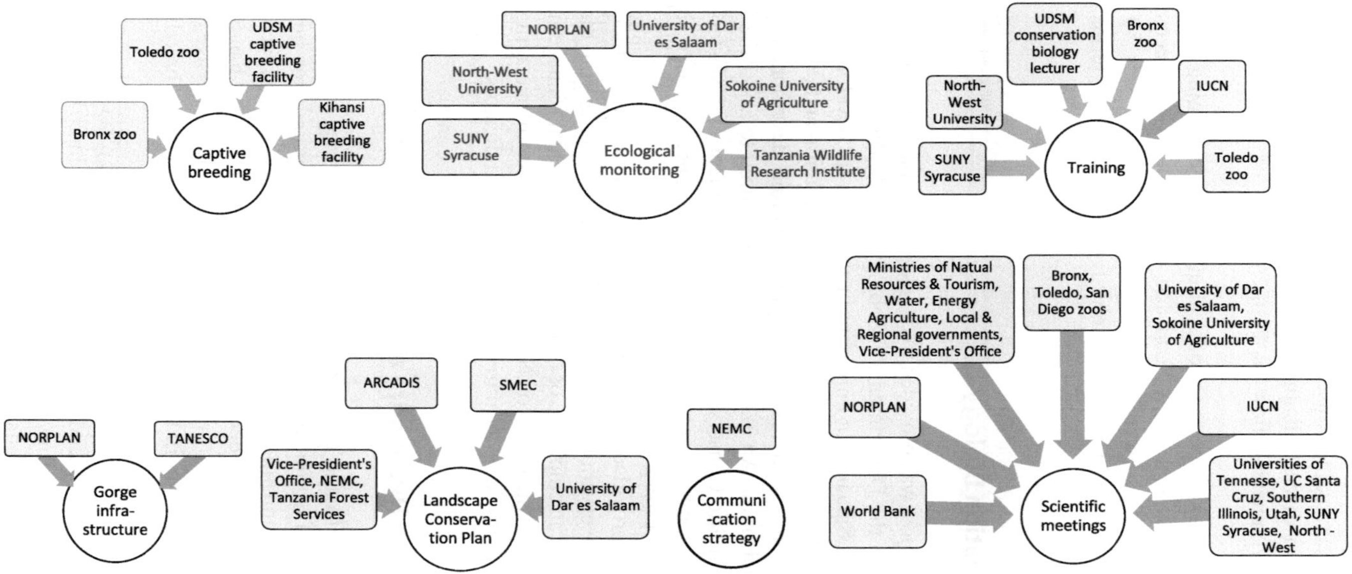

Fig. 7.1 Institutional inputs between 2001 and 2019 to the development of the major components (captive breeding, ecological monitoring, training, gorge infrastructure, landscape conservation plan, communication strategy, and scientific meetings) of the Kihansi Spray Toad recovery program

During the first decade of LKEMP (2002–2012), a training program was developed and implemented that focused on developing expertise within Tanzania in multiple disciplines including conservation biology, ecological monitoring, captive breeding, amphibian disease diagnostics, database management, water resource management, environmental flow assessment, and environmental policy and compliance.

Because of the critical importance of institutional strengthening to the long-term conservation of the Kihansi Spray Toad and the Kihansi Gorge ecosystem—the designed lifetime of the Lower Kihansi hydropower facility was 50 years—the training component under LKEMP received high priority in terms of project implementation.

7.3.1 Conservation Biology

Under LKEMP, support was provided to develop and establish a curriculum in conservation biology in the Department of Zoology and Wildlife Conservation at the University of Dar es Salaam. An important justification for developing this curriculum was the recognition that conflicts between conservation and development were increasing in Tanzania as a result of rapid economic development, population growth, and globalization. An external lecturer was hired to assist the Department of Zoology and Wildlife Conservation in the development of a Masters' level curriculum in conservation biology as well as provide lectures in conservation biology to undergraduate and post-graduate students.

Support was also provided for the overseas training of two MSc students and two Ph.D. students in conservation biology. Additionally, a scientist from the University of Dar es Salaam and an employee of National Environmental Management Council were trained at State University of New York, Syracuse, in managing ecological databases and analyzing long-term trends in ecological indicators.

7.3.2 Amphibian Husbandry and Disease Diagnostics

Between 2008 and 2013, the Toledo and Bronx zoos provided extensive training for national scientists and technicians in captive amphibian population husbandry and management, animal nutrition, disease diagnostics, pathology, and facility management. This training was provided in the form of short-courses and hands-on training at these two zoos. The Toledo and Bronx zoos also provided extensive on-site training and support in Tanzania in the establishment of two captive breeding facilities for the Kihansi Spray Toad.

As part of training in disease diagnostics and pathology, scientists from Sokoine University of Agriculture attended a month-long course in 2010 in amphibian disease diagnostics and pathology at the Bronx and San Diego zoos in the United States and in real-time PCR analysis for detection of chytrid fungus at North-West University in South Africa.

7.3.3 Amphibian Reintroductions

Between 2012 and 2014, two IUCN consultants were hired to initiate and advise the Tanzanian government on the reintroduction of the Kihansi Spray Toad back to the wild and to co-advise a MSc student at the University of Dar es Salaam in amphibian reintroduction and population biology. Support was also provided between 2012 and 2014 for a second MSc student from Sokoine University of Agriculture to assess the epidemiology of chytrid fungus in amphibians in the Udzungwa Mountains. This student was advised by the senior scientist at Sokoine University of Agriculture, who had been trained in real-time PCR analysis of chytrid fungus at North-West University in South Africa.

7.3.4 Water Resource and Environmental Management

In 2001 there was also little capacity within and coordination among government agencies responsible for managing the environment and water resources in Tanzania. Thus, an important initial activity under LKEMP was to develop institutional capacity to address multi-sectoral environmental and water resource development challenges.

LKEMP also provided support for in-service training of personnel in the National Environmental Management Council (NEMC) and Department of Energy (DOE) in conducting environmental impact assessments as well as support for the development of an environmental unit within the Ministry of Energy and Minerals (MEM). Support was also provided to the University of Dar es Salaam to enhance the curriculum in water resource management and specifically in the assessment and management of environmental flows.

Finally, under the Kihansi Catchment and Conservation Project (KCCMP), a follow-on project to LKEMP, support was provided to train district and village environmental officers, extension workers, and community leaders in the Kihansi catchment in integrated pest management and in promoting land use practices that enhance water quality and quantity in the catchment. As part of the village-focused training, villagers were involved in the mapping and demarcation of all water sources in the Kihansi catchment.

Reference

Hirji R (2006) Annex 7. Lower Kihansi Environmental Management Project (LKEMP). Implementation review mission

Implementation of the Lower Kihansi Environmental Management Project, 2002–2003

8

Abstract

Three major activities that were incorporated into the species recovery program for the Kihansi Spray Toad were initiated under the Lower Kihansi Environmental Management Project between 2002 and 2003. These included (1) convening an external panel to review past research and monitoring and to recommend future research and monitoring; (2) development of monitoring protocols; and (3) assessing the response of the Kihansi Spray Toad and the spray wetland habitat to the installation of the spray irrigation systems in the Upper, Lower, and Mid-Gorge spray wetlands. The population of the Kihansi Spray Toad in the Kihansi Gorge rebounded between October 2000 and June 2003. By early June 2003, the population of the Kihansi Spray Toad in the Kihansi Gorge was 20,990, very likely the size of the population in the Kihansi Gorge prior to the commissioning of the Lower Kihansi hydropower facility.

In July 2001, the Board of Governors of the World Bank approved the 5-year $6.1 million Lower Kihansi Environmental Management Project (LKEMP). Although the development of LKEMP was fast-tracked within the World Bank—it was designed and approved in less 6 months—the project did not commence until April 2002.

The objective of LKEMP was to put in place a series of measures for the long-term conservation of the Kihansi Gorge ecosystem. This objective was to be achieved through the implementation of a series of medium-term measures. The project comprised of four components:

- Habitat and species conservation and management.
- Establishment of a final water right.

W. D. Newmark, *The Kihansi Spray Toad* (Nectophrynoides asperginis): *Crisis Biology in the 21st Century*, Ecological Studies 254, https://doi.org/10.1007/978-3-031-89599-9_8

- Implementing an updated environmental management plan.
- Institutional strengthening.

Between 2002 and 2003, three major activities were initiated under the first component, habitat and species conservation and management: (1) convening an external panel to review past research and recommend future research and monitoring; (2) establishing an ecological monitoring program for the Kihansi Gorge; and (3) assessing the population response of the Kihansi Spray Toad to the installation of the spray irrigation systems and the effectiveness of the spray irrigation system in maintaining the spray wetlands.

8.1 External Review Panel, 2003

Most of the research that had been conducted within the Kihansi Gorge ecosystem up until 2001 had been focused on identifying the threatened and endemic plant and animal species that were impacted by the Lower Kihansi hydropower facility. The next step was to begin to understand the population dynamics and trends in the plant and animal species that were restricted to the spray wetlands and the effectiveness of the mitigation measures that were initiated under the IREM project to conserve the Kihansi Gorge ecosystem.

At the request of the Tanzanian government, LKEMP convened in January 2003 an external panel to review the ecological research and monitoring that had been conducted in the Kihansi Gorge since 1995 and to recommend future research and monitoring to fill important gaps in knowledge. This panel consisted of Daniel Doak (University of California, Santa Cruz), Louis Gross (University of Tennessee), James Gibbs (State University of New York, Syracuse), Karen Lips (University of Southern Illinois), and Diane Davidson (University of Utah).

Each of the panelists were tasked with reviewing past research and monitoring within one or more of the following areas: microclimate, amphibians, entomology, vegetation, intermittent bypass flows, and translocation feasibility. The external review panel's Tanzanian counterparts included Raphael Mollel, Permanent Secretary in the Vice President's Office; Magnus Ngoile, Director General of the National Environment Management Council; Idris Kikula, University of Dar es Salaam; and Wilfred Sarunday, project coordinator for LKEMP. Jan Lindemark and Svein Haarklau represented NORPLAN, and the World Bank delegation included Ladisy Chengula, task team leader for LKEMP, and myself, a consultant to the World Bank.

The panelists were impressed with the quality of the research that had been conducted between 1995 and 2002 in the Kihansi Gorge under the NORPLAN Lower Kihansi Long Term Environmental Monitoring Program and IREM and strongly supported the continuation of the ongoing studies and monitoring.

The panel proposed the following objectives for ecological monitoring in the Kihansi Gorge:

1. Assess the long-term impact of water diversion on the Kihansi Gorge ecosystem.
2. Evaluate the effectiveness of management measures to maintain species diversity and ecological processes in the Kihansi Gorge.
3. Provide knowledge about the structure and function of the Kihansi Gorge ecosystem for purposes of designing effective mitigation measures.

The panel also proposed that the objectives for future research in the Kihansi Gorge:

1. Provide critical information and knowledge necessary to develop effective mitigation measures for the conservation of the Kihansi Gorge ecosystem.
2. Provide critical information and knowledge necessary to understand the structure and function of the Kihansi Gorge ecosystem so as to enhance its long-term conservation.

The panel's taxon-specific recommendations for future monitoring and research are presented in Table 8.1.

Finally, the general policy recommendations of the External Review Panel were:

1. A program should be initiated immediately to reduce the likelihood of the accidental introduction of disease that could threaten the Kihansi Spray Toad by more carefully controlling visitation to the gorge and by requiring all individuals who travel into the gorge sterilize their footwear and any equipment they may be carrying. In addition, fish should not be introduced into the reservoir because of the risk of accidental interspecific transmission of disease. Finally, careful attention needs to be given to the management of the upstream catchment so as to maintain water quality in the gorge.
2. Data compilation and analysis should be conducted on a monthly basis to ensure that emerging answers from the data can be used to influence mitigation procedures. Otherwise, monitoring will not feedback into management of the Kihansi Gorge.
3. LKEMP should identify an individual, preferably the field coordinator, who will be responsible for managing and backing up the monitoring and research databases. As part of this database management, data need to be compiled and stored in a fashion that will allow easy and rapid access and analysis.
4. Conditions should be attached to the final water that specify that TANESCO will exercise "due diligence" to conserve the Kihansi Gorge ecosystem and that the health of the Kihansi Gorge ecosystem will be assessed every 3–5 years by an independent scientific panel that will have the authority to require further mitigation measures and/or modify existing mitigation practices.

Table 8.1 Taxon-specific recommendations for monitoring and research proposed by the External Review Panel in January 2003

Taxa	Recommendation for monitoring and research
Kihansi Spray Toad	1. Determine whether PVC pipes placed in permanent positions can be used to more easily and powerfully sample Kihansi Spray Toad numbers in the wetland vegetation. The methodology has been used previously to monitor frog populations in other areas
	2. Syringe irrigation of live toads to acquire diet samples. These samples can be used to determine what the toads eat, including seasonal changes in food items throughout the year, as well as differences in food use among age classes
	3. Conduct a mark-resight study to establish more accurately survival rates of the Kihansi Spray Toad. Colored elastomer markers could be used to mark individuals. The resulting data could be used to examine whether collecting individuals to assess reproduction and disease rates should be done
	4. Use repeated counts over short time intervals so that true observation error can be estimated
	5. Determine whether auditory surveys with automatic recording equipment can sample Kihansi Spray Toad populations
	6. If there is evidence that it will not harm Kihansi Spray Toad populations, conduct repeated harvesting of relatively small numbers of toads from wild populations to collect reproduction and disease rate data
	7. Conduct genetic analyses to establish both the genetic diversity of each of the Kihansi Spray Toad populations and also the level of genetic differentiation between populations
Insects	1. Sample flying insect abundance to identify seasonal or irregular emergence events (mass emergence of insects with either aquatic or nonaquatic larvae). Such events are often triggered by rainfall. If there are big hatching events, they may provide a disproportionate amount of the Kihansi Spray Toad food needs and serve as a trigger for toad reproduction
Spray wetland vegetation	1. Increase the density and/or change the distribution of sprinklers to better recreate natural precipitation and soil moisture levels. Experimentation with water jets, soaker hoses, or drip irrigation could also be used in conjunction with sprinklers and jets. Data could be collected on responses of soil characteristics, plants, insects, and toads.
	2. Initiate experimental restoration of wetland plants. Two pieces to such experiments are weeding out of species not believed to be natural components of the wetland flora and planting of formerly dominant wetland species that are not yet showing clear recovery (e.g., *Selaginella*). Both the timing and intensity of the weeding and planting should be varied to determine the maximally effect way to encourage return to more natural wetland communities
	3. Conduct stable isotope and nutrient analyses of plant samples from the wetlands to look for difference of nutrient and stress levels both between wetlands and resulting from the sprinkler systems
	4. Map and follow individuals of the rare and endemic plant species in the wetlands. With periodic resurveys, mortality and recruitment rates of these unique plants could be estimated, as well their responses to the sprinklers and other mitigation measures

(continued)

Table 8.1 (continued)

Taxa	Recommendation for monitoring and research
Riverine forest	1. Conduct stable isotope analysis and also tree ring analyses of trees in permanent forest plots. If these trees preserve annual growth rings, the width of these rings could be compared from before diversion to the current growth, providing a powerful way to gauge diversion effects on the riverine forest. Isotopic analysis will provide a second way to estimate physiological (water) stress on forest trees
	2. Preliminary work to establish a cost-effective method with which to sample epiphytes in the riverine forest. Most probably, data on numbers, diversity, and physiological status of epiphytes could be sampled on the marked trees in the permanent woody plant plots already established. The logic of sampling epiphytes rests on their sensitivity to precipitation and humidity in the forest, thus making these plants excellent indicators of any widespread effects of water diversion from the gorge
Climatic and other abiotic factors	1. Quantify the abiotic variables (means and variances) of the spray wetlands with stable or increasing versus declining or declined toad populations
	2. Addition of silt to wetland areas to gauge the effects of former inputs from natural spray. Since sprinkler water is filtered, wetlands are no longer receiving natural levels of waterborne materials, with possible effects on wetland plants
	3. Additional intermittent flow release experiments, perhaps for longer periods of time, to better measure effects on edges of wetlands
Laboratory experiments	1. Measure length-weight relationships so that weight of wild toads can be used to determine reproductive status of females
	2. Conduct habitat preference and performance experiments of Kihansi Spray Toad, with particular reference to wind, spray, vegetation, and humidity. However, this type of study might be less expensive and more accurate if done in the field
	3. Test desiccation resistance of plants

8.2 Development of Monitoring Protocols

One of the more important initial priorities of LKEMP and which was implemented during the first 3 years of the project was to develop monitoring protocols to assess the effectiveness of mitigation measures developed under the LKHP and IREM projects to conserve plant and animal communities in the Kihansi Gorge.

Under LKEMP, it was envisioned that ecological monitoring would be integrated into an adaptive management framework that would guide project design and implementation of activities to conserve the Kihansi Spray Toad and its habitat. It was also envisioned that the monitoring of key indicators would allow a rapid assessment of short- to medium-term project objectives and provide evidence for the continuation or change of project activities to conserve the Kihansi Spray Toad and its habitat. Potential indicators would be prioritized for inclusion in a monitoring program based on their information content and ecological relevance, measurement costs, technical feasibility, and quality of baseline data.

Fig. 8.1 James Gibbs served on the External Review Panel, designed the ecological monitoring protocols for the Kihansi Gorge, and conducted toxicology experiments on the Kihansi Spray Toad under LKEMP. (Photo credit: James Gibbs)

In August 2003, LKEMP contacted James Gibbs (Fig. 8.1), a member of the external review panel and an expert in ecological monitoring, and requested his assistance in designing monitoring protocols for the Kihansi Gorge. Over the next 2 years, James Gibbs developed detailed monitoring protocols for plant and animal species and environmental parameters in the Kihansi Gorge. In developing these protocols, he reviewed both the strengths and limitations of the ecological protocols that were developed under LKHP and IREM and evaluated each of the potential protocols based on four criteria: (1) relevance to monitoring the effects of hydrological manipulation on the Kihansi Gorge ecosystem; (2) cost-efficiency in terms of the amount of monitoring information gained in relation to the effort expended to obtain it; (3) feasibility in terms of the degree of specialized technical expertise required to obtain the data; and (4) strength of pre-project data baseline (Table 8.2) (Gibbs 2006).

James Gibbs also conducted a statistical power analysis using data collected under LKHP and IREM to examine the likelihood of detecting changes in environmental indicators (Table 8.3). The value of the power analysis is that tradeoffs between effort expended and likelihood of successfully tracking the indicator can be examined.

This statistical power analysis specifically asked: "Does the sampling regime implemented by NORPLAN under the Long Term Ecological Monitoring Program and IREM have reasonable power (>80%) of detecting a 50% decline in the indicator's attribute between successive sampling episodes while generating false positives at a low level ($\alpha = 0.10$) (Gibbs 2006)?" Although there are multiple methods to assess power, a paired-sample t-test approach was selected because (1) it conformed to the type of sampling that had been occurring at Kihansi—permanent plots with comparisons focusing on changes between successive sampling

Table 8.2 A summary of the relationship of potential monitoring indicators for the Kihansi Gorge ecosystem relative to four criteria for inclusion in a final monitoring program: (1) relevance to the monitoring the effects of hydrological manipulation of the Kihansi River; (2) cost-efficiency in terms of the amount of monitoring information gained in relation to the effort expended to obtain it; (3) feasibility in terms the degree of specialized technical expertise required to obtain the data; and (4) strength of pre-project data baseline and other issues of data quality

Indicator	Relevance	Cost-efficiency	Technical feasibility	Strength of baseline	Score (max. = 12)
Spray wetlands monitoring	*High* Regionally significant ecosystem type directly impacted by river	*Intermediate* Intensive sampling but in restricted areas	*Intermediate* Expert botanical skills needed Plant identification guide already created by NORPLAN Species richness not excessive	*High* Extensive baseline data on permanent plots	10
Woody plants	*High* Woody plants represent most of biomass in Gorge Tree growth affected by moisture levels Some rare species present	*Low* Sampling involves intensive effort at many plots in difficult locations Data management and analysis issues substantial	*Low* High species richness of forest requires extensive expert knowledge Specialized mensuration techniques required Requires familiarity with statistical analysis and management of complex botanical data sets	*High* Extensive baseline data on permanent plots	8

(continued)

Table 8.2 (continued)

Indicator	Relevance	Cost-efficiency	Technical feasibility	Strength of baseline	Score (max. = 12)
Herbaceous plants	*Low* Few species of conservation concern demonstrated to occur during initial on surveys in permanent monitoring plots	*Low* Sampling involves intensive effort at many plots in difficult locations Data management and analysis issues substantial	*Low* High species richness of herbaceous community requires extensive expert knowledge Specialized mensuration techniques required Requires familiarity with statistical analysis and management of complex botanical data sets	*High* Extensive baseline data on permanent plots	6
Primates	*High* Group charismatic with public Includes two rare endemics	*Intermediate* Sampling involves extensive linear transects through difficult terrain (along gorge contours)	*Intermediate* Sampling involves some expert knowledge of primate identification Requires training in specialized survey methodology	*High* Extensive baseline data on transects that can likely be relocated	10
Small vertebrates	*Low* Few species of *demonstrated* conservation concern (too poorly known taxonomically in many cases)	*Low* Trapping grids and pitfall traps very labor intensive	*Low* Major component group—shrews—largely unidentifiable even by experts Expert identification an issue for other, non-shrew groups as well	*Low* High variability and low capture rates characterize baseline Relatively few baseline measurement sites owing to intensity of sampling effort	4

Avifauna	*Intermediate* Charismatic group Some rare forms present Some rare forms with significant populations in Gorge All rare forms found elsewhere in E. Arc Mountains	*Low* Mist netting in difficult to reach locations Mist netting labor intensive	*Intermediate* Requires knowledge of ringing methods Requires knowledge of mist netting practices Requires knowledge of bird identification	*High* Baseline well established for permanent plots	8
Invertebrates—*Afrosteles distans*	*Intermediate* Useful invertebrate indicator of spray wetland quality May indicate waterborne insecticide exposure Minor component of toad diets Not of particular conservation concern although possibly a locally adapted form	*Intermediate* Intensive sampling methods in restricted area Post-processing of samples in laboratory required	*Intermediate* Requires some specialized knowledge of invertebrate identification	*High* Baseline well established for permanent plots	9

(continued)

Table 8.2 (continued)

Indicator	Relevance	Cost-efficiency	Technical feasibility	Strength of baseline	Score (max. = 12)
Invertebrates—*Ortheziola* sp.	*Intermediate* Possible invertebrate indicator of spray wetland quality May indicate waterborne insecticide exposure Not a particularly important component of toad diets May represent new species restricted to gorge	*Intermediate* Intensive sampling methods in restricted area Post-processing of samples later in laboratory (sorting and counting)	*Intermediate* Requires some specialized knowledge of invertebrate identification	*Intermediate* Baseline established for permanent plots Evolving improvements to sampling techniques present issues for data comparability among surveys	8
Kihansi Spray Toad	*High* Endemic species restricted to gorge Physiology and behavior intimately linked to changes in river hydrology	*Intermediate* Intensive sampling methods but within highly restricted areas	*High* Requires limited training to identify toads Requires limited training to undertake simple counting techniques	*Intermediate* Baseline well established for permanent plots and transects Some inconsistency in sampling sites/transect locations complicate comparisons	10

Fisheries	*High* Provides vital perspective on landscape level changes within catchment that can influence many biota and physical processes within Gorge	*Low* Labor intensive surveys Specialized equipment Large field crews	*Low* Many undescribed or poorly known species requiring expert identification skills	*Intermediate* Baseline established but estimates highly variable Control site [Udugaji] likely invalid	6
Land cover	*High* Direct measurements of physical characteristics of hydrological environment Vital to tracking change in affected resource Vital to interpreting changes in other resources	*Intermediate-low* Depends on technology used: *Aerial photography* Straightforward methodology High information content Laborious to classify and transfer to maps Requires extensive field verification of classifications *Remote digital imagery* Expensive to acquire, cheap to process if technical skills available to do so Requires extensive field verification of classifications	*Intermediate-low* *Remote digital imagery* Technically demanding to process *Aerial photography* Requires photogrammetry skills Requires expert knowledge of local ecosystems to map them effectively	*Intermediate* No baseline exists Knowledge of ongoing changes would still inform management Remote imagery possibly available for pre-project conditions	8

(continued)

Table 8.2 (continued)

Indicator	Relevance	Cost-efficiency	Technical feasibility	Strength of baseline	Score (max. = 12)
River flow and quality (discharge, water pH, conductivity, temperature, suspended solids)	*High* Direct measurements of physical characteristics of sprinkler system water of great importance to viability of toad population	*High* All measurements can be taken at regular intervals at low cost per measurement once required recording equipment is secured	*High* Measurement of all parameters not highly technical	*High* Useable baseline data for most parameters exists in some form	12
Water quality associated with sprinkler system (flows, temperature, pH, conductivity, suspended solids)	*High* Direct measurement of physical variables directly responsible for vulnerable river-associated ecosystems	*High* All measurements can be taken at regular intervals at low cost per measurement once specialized required recording equipment is secured	*High* Measurement of most parameters straightforward although some technical ability required to operate automated devices	*High* Useable baseline data for most parameters exists	12
Climate and microclimate in spray wetlands (humidity, temperature, precipitation, wind speed)	*High* Direct measurement of important physical variables influencing ecosystems in gorge area	*Intermediate* Sophisticated automated recording devices provide opportunity to measure parameters continuously over small area	*Intermediate* Some technical ability required to operate automated devices	*High* Useable baseline data for most parameters exists	10
Climate within Gorge (humidity, temperature, precipitation)	*High* Direct measurement of important physical variables influencing ecosystems in gorge area	*High* Automated recording devices [microloggers] provide opportunity to measure parameters over large areas cheaply	*Intermediate* Some technical ability required to operate and deploy automated devices	*High* Useable baseline data for most parameters exists	11

The status of each indicator relative to each criterion is displayed in terms of three qualitative assessments: "high," "intermediate," and "low." The overall score is the sum of component scores, where high is 3, intermediate is 2, and low is 1. Source: Gibbs (2006)

Table 8.3 Power analysis of potential environmental indicators for the Kihansi Gorge ecosystem

Indicator	Attribute	Mean	Standard deviation of estimates	Standard deviation of differences between estimates	Current plots/ transects	Power to detect 50% decline	Estimated plots/transects needed to detect 50% decline at ($\alpha = 0.10$)
Primates	Sykes Monkey (individuals)	4.3	6.18	5.4	8	0.27	40
	Red Colobus (individuals)	11.8	11.25	14.0	8	0.22	51
	Black and white Colobus (individuals)	2.6	5.09	7.2	8	0.14	>100
	All species (individuals)	19.5	16.72	19.4	8	0.36	26
Woody plants	Basal area growth (%)	0.0028	0.0019	0.00177	28	1.00	3
Birds	All species (individuals)	14.6	6.33	7.7	4	0.41	9
	Species richness	6.3	2.35	2.6	4	0.60	6
Fisheries	Catch/effort	76.1	89.73	160.3	4	0.12	>100
Small vertebrates	Small mammals (individuals/100 trap nights)	10	5.75	7.4	3	0.21	16
	Amphibians (individuals/100 trap nights)	40.9	58.62	90.2	3	0.11	>100
	Reptiles (individuals/100 trap nights)	0.017	0.015	0.011	3	0.25	12
Invertebrates	*Afrosteles* (individuals/ sample)	9.1	12.98	12.1	8	0.23	51
	Ortheziola (individuals/ sample)	3.4	8.92	4.4	8	0.26	43
Kihansi toads[a]	Density (individuals/m^2)	~2.5	~2.77	~1.8	~8	~0.55	~15

(continued)

Table 8.3 (continued)

Indicator	Attribute	Mean	Standard deviation of estimates	Standard deviation of differences between estimates	Current plots/ transects	Power to detect 50% decline	Estimated plots/transects needed to detect 50% decline at ($\alpha = 0.10$)
Kihansi toads[a]	Density (individuals/m^2)	~2.5	~2.77	~1.8	~180	~0.55	~60
Wetland vegetation	*Microstegium vagans* (%)	46	20.1	26.6	10	0.80	10
	Selaginella kraussiana (%)	7	5.32	5.7	10	0.55	14
	Brillantaisia madagascariensis (%)	23.4	19.52	19.8	10	0.53	19

Mean equals average value detected during pilot studies conducted by NORPLAN. Standard deviation of estimates equals that observed across all plots/transects (on detrended data). Standard deviation of differences between estimates equals that observed for differences of later minus later estimates on the same plots or transects. Current plots/transects equal the sampling effort deployed during NORPLAN monitoring. Power to detect 50% equals the likelihood of detecting a halving of the mean value based on a paired-sample t-test. The last column is the estimated number of plots or transects needed to detect the halving of the observed mean with a likelihood of 80% while wrongly concluding a trend occurred 10% of the time. Source: (Gibbs 2006)

[a]Uncertainties concerning estimates of spray toad parameters relate to varying numbers of plots and transect locations used to survey toads on different occasions

Table 8.4 Prioritized list of indicators for inclusion in the ecological monitoring program at the Kihansi Gorge as identified by natural resource managers in Tanzania

Tier	Ecological indicator	Relevance	Cost efficiency	Technical feasibility	Strength of baseline	Total
First tier						
1	Flow and water quality (discharge, water pH, conductivity, temperature, suspended solids) in main stem of the river	2.8	2.2	2.6	3	10.6
2	Kihansi spray toad along with other small vertebrates including amphibians of the spray wetlands	2.8	2.3	2.5	3	10.6
3	Climate and microclimate in spray wetlands (humidity, temperature, precipitation, wind speed)	2.8	2.2	2.4	3	10.4
Second tier						
4	Woody plants	2.5	2.3	2.3	3	10.0
5	Spray wetland plant communities	2.5	2.2	2.2	3	9.9
6	Climate within Gorge (humidity, temperature, precipitation)	2.6	2.0	2.1	3	9.7
7	Spray wetland invertebrates *Afrosteles distans* and *Ortheziola* sp.	2.4	2.1	2.2	3	9.6
8	Spray wetland insect community in general	2.4	2.1	2.1	3	9.6
9	Water quality associated with sprinkler system (flows, temperature, pH, conductivity, suspended solids)	2.8	2.3	2.4	2	9.5
Third tier						
10	Avifauna	2.3	1.7	2.1	3	9.1
11	Land cover	2.5	2.2	2.3	2	9.0
12	Primates	2.2	1.8	1.9	3	8.9
Fourth tier						
13	Flow and water quality (discharge, water pH, conductivity, temperature, suspended solids) in streams of the upper catchment	2.7	2.3	2.4	1	8.4
14	Small vertebrates of the forest	1.9	1.8	1.7	3	8.4
15	*Coffea kihansiensis*	2.8	2.1	2.4	1	8.3

(continued)

Table 8.4 (continued)

Tier	Ecological indicator	Relevance	Cost efficiency	Technical feasibility	Strength of baseline	Total
16	Chytrid fungi of amphibians	2.6	2.4	2.2	1	8.2
17	Flow and water quality in streams feeding the sprinkler systems	2.6	2.2	2.3	1	8.1
18	Fish diversity in the main river	2.4	1.6	2.0	2	8.0
19	Fisheries	2.4	1.8	1.8	2	8.0
Fifth tier						
20	Management system	2.5	2.2	2.0	1	7.6
21	Reptiles in forest	1.9	1.8	1.9	2	7.6
22	Pollutants (pesticides, metals) in Kihansi River	2.4	2.1	2.1	1	7.6
23	Oxygen levels in river	2.3	2.2	2.2	1	7.6
24	Occurrence of fires (direct observation from gorge)	2.3	2.2	1.9	1	7.5
25	Epiphylls	2.3	2.1	2.1	1	7.5
26	Soil quality	2.3	2.1	2.0	1	7.3
27	Herbaceous plant community	2.3	1.7	2.2	1	7.2
28	Stream macroinvertebrates, particularly dragonfly larvae	2.4	1.9	1.9	1	7.1
29	Forest butterfly communities	2.2	1.8	2.1	1	7.1
30	Predators of Kihansi spray toad	2.4	1.8	1.8	1	7.1
Sixth tier						
32	Red ants	2.0	1.8	1.7	1	6.4
33	Earthworms	1.6	1.8	1.8	1	6.2

The numerical value within a column is the average evaluation score generated by workshop participants across the sum of the individual scores for each evaluation criterion: (1) *relevance* to monitoring the potential ecological effects of hydrological manipulation of the Kihansi River; (2) *cost-efficiency* in terms of the amount of monitoring information gained in relation to the sampling effort needed to obtain it; (3) *feasibility* in terms of the degree of specialized technical expertise required to obtain the data; and (4) *strength* of pre-project data baseline (high = 3, intermediate = 2, and low = 1). Natural breaks in the average scores were used to identify different tiers of priority for indicators to include in the monitoring program. Source: Gibbs (2006)

intervals—and (2) the importance of providing to management authorities rapid detection of environmental change between successive sampling episodes.

In early January 2006, LKEMP invited 28 natural resource managers from throughout Tanzania to finalize the selection of a set of ecological indicators to be monitored in the Kihansi Gorge. Ecological indicators were ranked into six tiers (Table 8.4). The top-two tiers became the primary focus of ecological monitoring within the Kihansi Gorge over the next 13 years.

Finally, James Gibbs and Matthew Buff, a graduate student in Gibbs' lab, developed a data management system for the ecological monitoring data. The data management system was designed not only to allow data importation and storage but also to assess trends in ecological indicators and reporting and information dissemination. Two Tanzanian scientists were trained to manage the data management system.

Between 2006 and 2019, consultants from the Department of Zoology and Wildlife Conservation and Department of Botany at the University of Dar es Salaam were contracted by LKEMP and later by KCCMP to monitor the first- and second-tier ecological indicators in the Kihansi Gorge. The monitoring of the first- and second-tier ecological indicators was conducted at intervals of 1–4 years largely following the agreed upon methodology. However, because of budgetary constraints in hiring a database manager and data entry personnel as well as the difficulty in extracting and formatting raw data from the consultants' reports in a form suitable for data entry, the ecological monitoring database established under LKEMP was never fully utilized at a level commensurate with its capacity and design.

8.3 Population Response of the Kihansi Spray Toad to the Spray Irrigation System

Assessing the population response of the Kihansi Spray Toad to the installation of the spray irrigation systems in the Mid-Gorge, Lower, and Upper Spray Wetlands was one of the highest priority activities implemented during the first 3 years of the LKEMP.

Between June 2000 and March 2001, NORPLAN had installed an artificial spray irrigation system within three of the five spray wetlands in the Kihansi Gorge: Mid-Gorge, Lower, and Upper. The Main Falls and Mhalala Spray Wetlands, the two smallest spray wetlands, were maintained by natural spray from adjacent waterfalls.

Although the long-term impact of the installation of the spray irrigation system on the population of the Kihansi Spray Toad in the Kihansi Gorge was positive, there was a short-term lag in the response.

Between October 2000 and March 2001, following the commissioning of the Lower Kihansi hydropower facility and prior to completion of the installation of the spray irrigation system in the Lower and Upper Spray Wetlands, the Kihansi Spray Toad population in the gorge declined from approximately 12,100 individuals in the gorge to 1412 (Fig. 8.2). Over this time period, nearly all of the Kihansi Spray Toads

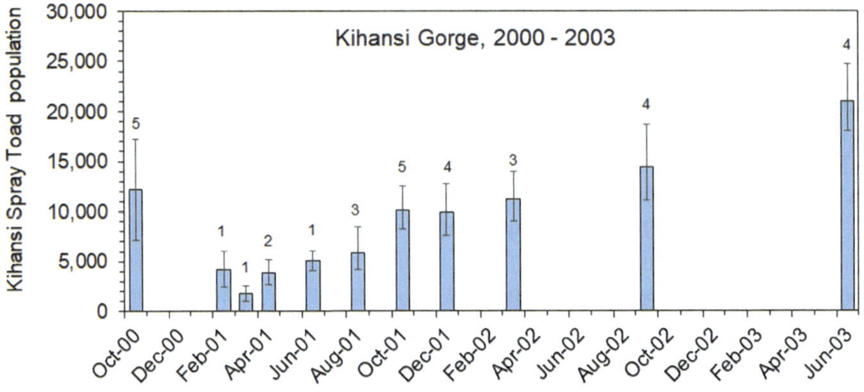

Fig. 8.2 Change in the population of the Kihansi Spray Toad in the Kihansi Gorge between October 2000 and June 2003. Error bars represent the 95% confidence intervals of the Kihansi Spray Toad population in the spray wetland vegetation based on bias-corrected and accelerated (BCa) bootstrap intervals. The Kihansi Spray Toad population on rock faces in the Kihansi Gorge was estimated from total counts. The number of wetlands within the Kihansi Gorge that were included in each survey is shown above the population estimate. Only the Upper Spray Wetland, the largest wetland in the Kihansi Gorge, was included in all surveys. (Source: Newmark 2002; NORPLAN 2003)

in the Upper, Lower, and Mid-Gorge spray wetlands were concentrated on wet rock faces close to the Kihansi River.

Beginning in June 2001 with the completion of the installation of the spray irrigation systems in the Upper, Lower, and Mid-Gorge Spray Wetlands, the Kihansi Spray Toad population in the Kihansi Gorge began to rebound and continued to expand over the following 2 years (Fig. 8.2). The spatial distribution of the spray toads within the Upper, Lower, and Mid-Gorge Spray Wetlands also shifted from wet rock faces along the Kihansi River into the adjacent spray wetland vegetation.

By early June 2003, the population of the Kihansi Spray Toad in the Kihansi Gorge had rebounded to 20,990 (Fig. 8.2), very likely the approximate population size of the Kihansi Spray Toad in Kihansi Gorge prior to the commissioning of the dam.

8.3.1 Population Response by Spray Wetland

The impressive recovery in the Kihansi Spray Toad population within the Kihansi Gorge between 2000 and 2003 was largely a result of the increase in the Kihansi Spray Toad's population in the Upper Spray Wetland (Fig. 8.3). The Upper Spray Wetland is the largest of the five spray wetlands and comprises 65% of the total area of the five spray wetlands within the gorge (Table 5.2).

In October 2000, shortly after the commissioning of the Lower Kihansi hydroelectric facility, the Upper Spray Wetland contained 89% of the estimated total population in the Kihansi Gorge (Table 5.2). Following the installation of the spray

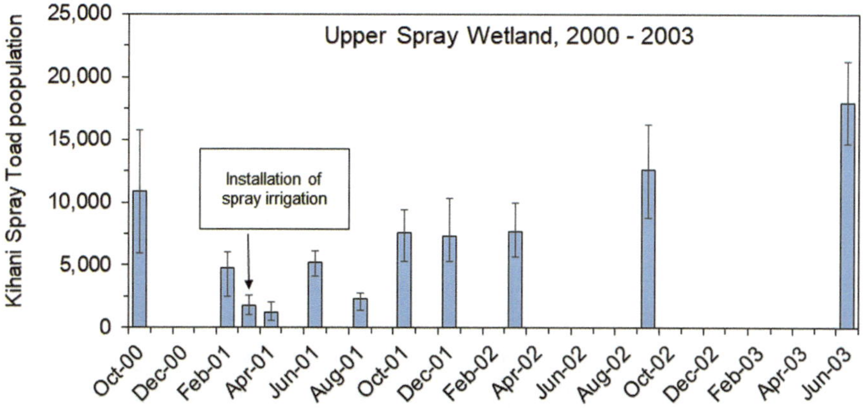

Fig. 8.3 Change in population of the Kihansi Spray Toad between October 2000 and June 2003 in the Upper Spray Wetland. Error bars represent 95% confidence intervals of the Kihansi Spray Toad population in spray wetland vegetation based on BCa bootstrapping. The Kihansi Spray Toad population on rock faces was based on total counts. (Source: Newmark 2002; NORPLAN 2003)

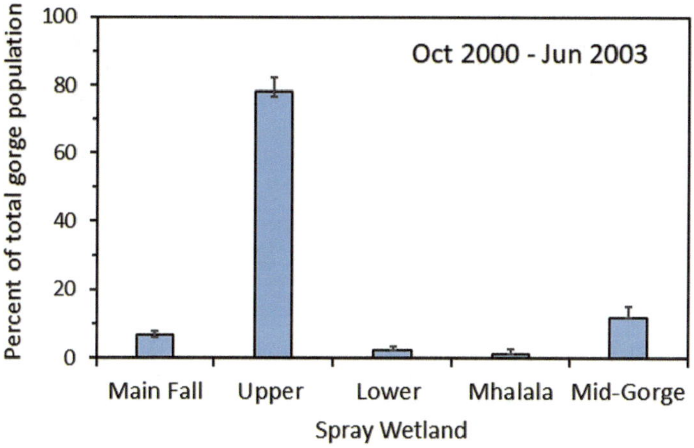

Fig. 8.4 Mean (± 1 SE) percent of the total Kihansi Spray Toad population in the Kihansi Gorge by spray wetland between October 2000 and June 2003. (Source: Newmark 2002; NORPLAN 2003)

irrigation systems in the Upper, Lower, and Mid-Gorge Spray Wetlands, the proportion of the total population of the Kihansi Spray Toad that was contained in the Upper Spray Wetland declined to 78% (range 67–89%) between October 2000 and June 2003 (Fig. 8.4).

Within the other four spray wetlands in the Kihansi Gorge, the population response of the Kihansi Spray Toad between October 2000 and June 2003 to the installation of the spray irrigation system varied by spray wetland. Within the Main Falls Spray Wetland, the only spray wetland along the Kihansi River that was maintained by the minimum bypass flow, the Kihansi Spray Toad initially population

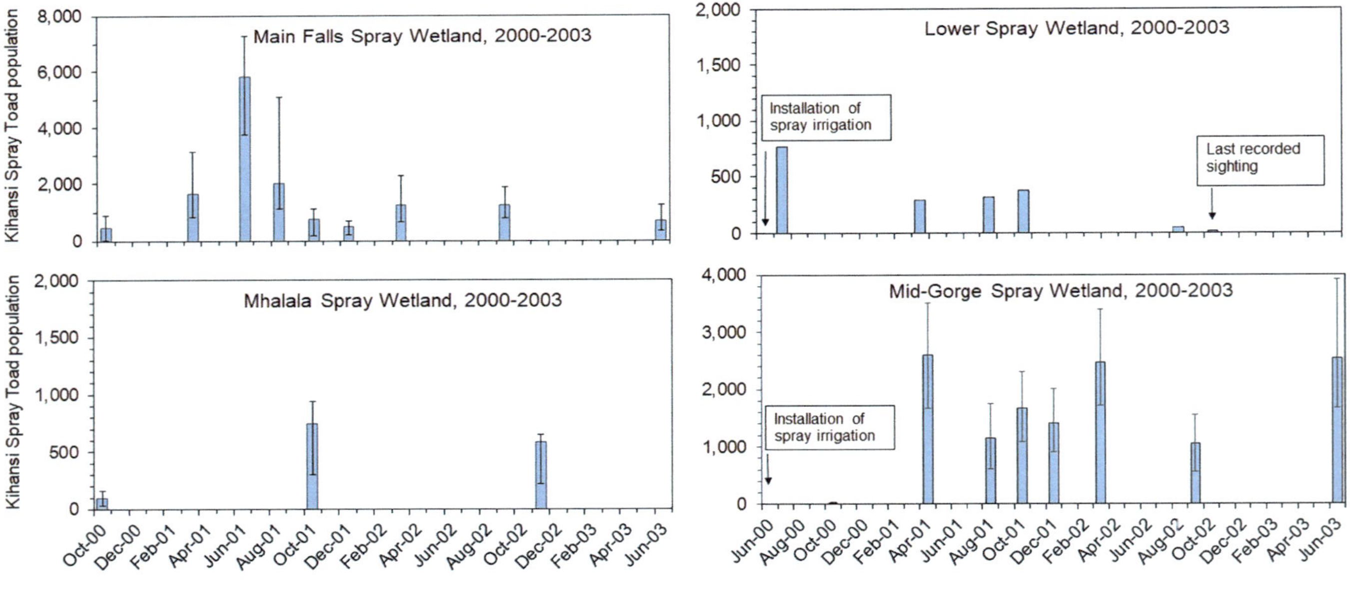

Fig. 8.5 Change in population of the Kihansi Spray Toad between October 2000 and June 2003 in the Main Falls, Lower, Mhalala, and Mid-Gorge spray wetlands in the Kihansi Gorge. Error bars represent 95% confidence intervals and are calculated from estimates of population size in spray wetland vegetation. Population estimates on rock faces were based on total counts. Errors bars are not presented in the Lower Spray Wetland after October 2000 because estimates of population size were too small to estimate confidence intervals. (Source: Newmark 2002; NORPLAN 2003)

grew and then declined very rapidly over this time period. In the Mid-Gorge and Mhalala Spray Wetlands, the spray toad populations remained relatively stable, while in the Lower Spray Wetland, the population steadily declined between 2000 and 2002 with the last sighting recorded in November 2002 (Fig. 8.5).

The Lower Spray Wetland is the second largest spray wetland in the Kihansi Gorge. In October 2000, the Lower Spray Wetland also contained the second largest population of the five wetlands in the Kihansi Gorge. However, following the installation of the spray irrigation system in the Lower Spray Wetland, the Kihansi Spray Toad population steadily declined with the last recorded sighting of a spray toad in late 2002 (Fig. 8.5). One hypothesis that was proposed for the steady decline of the Kihansi Spray Toad in the Lower Spray Wetland was the longer delay following the commissioning of the dam in installing the spray irrigation system in the Lower Spray Wetland than in the Mid-Gorge Spray wetland. The spray irrigation system in the Lower Spray Wetland was installed in September 2000 in comparison to June 2000 in the Mid-Gorge Spray Wetland. On the other hand, the spray irrigation system in the Upper Spray Wetland was installed 6 months after the installation of the spray irrigation system in the Lower Spray Wetland, yet the Kihansi Spray Toad population in the Upper Spray Wetland quickly recovered in this wetland (Fig. 8.3). Thus, the decline of the Kihansi Spray Toad population in the Lower Spray Wetland between October 2000 and June 2003 remains an enigma.

Yet notwithstanding the decline of the Kihansi Spray Toad in the Lower Spray Wetland, I remember thinking in early June 2003 that the spray irrigation system had conserved the Kihansi Spray Toad in the wild and that the objective of the LKEMP to put in place a series of measures for the long-term conservation of the Kihansi Gorge ecosystem had been largely achieved. Unfortunately, this elation in June 2003 rapidly changed to stunned disbelief over the following 3 weeks.

References

Gibbs JP (2006) A detailed ecological monitoring protocol for the lower Kihansi gorge ecosystem. Final report to the Lower Kihansi Environmental Management Project

Newmark WD (2002) Conserving biodiversity in East African forests: a study of the Eastern Arc Mountains. Springer

NORPLAN (2003) Intermittent high low manipulations and ecological studies. Final report. Report produced for the Lower Kihansi Environmental Management Project

Implementation of the Lower Kihansi Environmental Management Project, July 2003–2007

9

Abstract

In early July 2003, the spray irrigation maintenance team reported that they were no longer seeing any Kihansi Spray Toads in the Upper Spray Wetland. The decline began 2 weeks after an experimental intermittent bypass flow releases of 12 m^3/s and 16 m^3/s of water from the hydroelectric dam. A follow-up 1-hour nocturnal survey conducted in early August 2003 detected only 29 toads. This was shocking given that the population of the Kihansi Spray Toad in the Upper Spray Wetland in early June 2003 was approximately 18,000. A continuing and steady decline in the numbers of Kihansi Spray Toad was observed between June 2003 and May 2004 with the last recorded sightings of the Kihansi Spray Toad in May 2006. In 2009, IUCN formally listed the Kihansi Spray Toad as extinct-in-the-wild. Four prominent hypotheses have been proposed for the sudden decline and extinction in the wild of the Kihansi Spray Toad: (1) chytrid fungus *Batrachochytrium dendrobatidis*; (2) habitat alteration; (3) toxic contaminant; and (4) an interaction between *Bd* and a toxic contaminant. The best supported hypothesis is chytrid fungus. A retrospective analysis of 99 Kihansi Spray collected between 1986 and 2002 all tested negative for *Bd*; however five of eight specimens collected between August and November 2003 tested positive for *Bd*. Chytridiomycosis was also detected between 2004 and 2006 in two other anuran species occurring in spray wetlands in the Kihansi Gorge. The population recovery of the Kihansi Spray Toad in the Upper Spray Wetland following the installation of the spray irrigation system, and the undisturbed habitat in the Mhalala Spray Wetland where the Kihansi Spray Toad population also collapsed, provides countervailing evidence to the habitat alternation hypothesis. Experimental results on the toxicity of observed concentrations of endosulfan that were detected in reservoir sediments to the Kihansi Spray Toad were negative. A possible interaction between *Bd* and a toxic contaminant has not been evaluated.

W. D. Newmark, *The Kihansi Spray Toad* (Nectophrynoides asperginis): *Crisis Biology in the 21st Century*, Ecological Studies 254, https://doi.org/10.1007/978-3-031-89599-9_9

9.1 Collapse of the Kihansi Spray Toad Population

On 4 July 2003, the foreman of the spray irrigation maintenance team in the Kihansi Gorge informed LKEMP's project coordinator that the maintenance team was no longer seeing any Kihansi Spray Toads in the Upper Spray Wetland. He reported that the decline in spray toad numbers began during the second week in June approximately 1 week after the most recent survey of the Kihansi Spray Toad in the Upper Spray Wetland. During this survey that was conducted between the 3 and 4 June 2003, approximately 18,000 spray toads were recorded in the Upper Spray Wetland.

The maintenance team was instructed to continue to monitor and record the number of spray toads sighted during each of their visits to the Upper Spray Wetland. On 3 August 2003, Jeremy Thompson, LKEMP's technical advisor, conducted a 1-hour nocturnal survey of the Kihansi Spray Toad in the Upper Spray Wetland vegetation. During this nocturnal survey, he recorded one spray toad in 48 0.5 × 0.5 m plots in the Upper Spray Wetland spray wetland vegetation and 28 spray toads on the metal posts supporting the artificial spray irrigation system. The next day Jeremy Thompson conducted a half-hour diurnal survey in the Lower and Mid-Gorge Spray Wetlands. During these diurnal surveys, he did not observed any spray toads in the Lower Spray Wetland. In the Mid-Gorge Spray Wetland, he observed one juvenile toad and one dead adult, which he collected. Jeremy Thompson also visited the Mhalala Spray Wetland on 4 August and observed over a 15-minute period approximately 30 Kihansi Spray Toad on bare rock faces at the base of the waterfall. This latter observation confirmed the maintenance team's observation on 29 July 2003 of "many" Kihansi Spray Toads in the Mhalala Spray Wetland.

On 5 August 2003, I and two University of Dar es Salaam colleagues along with members from the Multi-sectoral Technical Advisory Committee (MTAC) joined Jeremey Thompson and conducted a diurnal survey of the Kihansi Spray Toad in the Main Falls, Upper, and Mid-Gorge Spray Wetlands. During these diurnal surveys, we observed one subadult in 0.5 observer-hours in the Main Falls spray wetland and one juvenile in 1.5 observer-hours in the Mid-Gorge Spray Wetland and did not observe any Kihansi Spray Toads in the Upper Spray Wetland in 5.0 observer-hours.

We believed based on Jeremey Thompson's August fourth observation of approximately 30 Kihansi Spray Toads on rock faces in the Mhalala spray that the population in this wetland may be stable. The last census of the Kihansi Spray Toad in the Mhalala survey had been conducted in November 2002 and had estimated a population of 582 animals of which 29 percent of the population had been observed on rock faces (NORPLAN 2002). Thus, we speculated that the comparative remoteness of the Mhalala Spray Wetland in relation to the other spray wetlands in combination with the lower visitation rate by maintenance staff and scientists had possibly protected the Kihansi Spray Toad population in this wetland. We, therefore, recommended that the maintenance staff refrain in the short term from visiting the Mhalala Spray Wetland.

Between 5 and 8 August, the maintenance team continued to conduct daily nocturnal surveys of the Kihansi Spray Toad in the Upper Spray Wetland. All of the toads that they recorded were observed on spray irrigation piping.

When Jeremy Thompson returned to Dar es Salaam on 6 August, he compiled a Fact Sheet on what was known regarding the sudden collapse of the Kihansi Spray Toad population in the Upper, Lower, and Mid-Gorge Spray Wetlands (Thompson 2003). The more likely causes that were proposed for the sudden collapse of the Kihansi Spray Toad population were disease and/or an accidental release of a toxicant associated with the experimental intermittent bypass flow releases that had been conducted on 31 May and 1 June 2003 for purposes of assessing the spatial dispersion of spray and the response of the Kihansi Spray Toad to these intermittent bypass releases. The experimental intermittent bypass release had occurred 3 days before the population in the wetland had been surveyed.

In January 2003, the external review panel had highlighted the threat that chytrid fungus *Batrachochytrium dendrobatidis* potentially posed to the Kihansi Spray Toad due to the recently reported declines of amphibians in Australia and Central America in the late 1990s (Berger et al. 1998). However, the panel's recommendation that all individuals entering any of the spray wetlands should sterilize their footwear and equipment both before and after entering a wetland had not yet been implemented.

In early August 2003, plastic footbaths to sterilize footwear were placed at the entrance of the Upper, Lower, Mhalala, and Mid-Gorge Spray Wetlands as well along the main foot path at the top and bottom of the Kihansi Gorge (Fig. 9.1). In 2016 the plastic footbaths were replaced with concrete footbaths (Fig. 9.2) along

Fig. 9.1 Plastic footbaths were installed at the entrance to all of the spray wetlands in the Kihansi Gorge and at the top and the bottom of the Kihansi Gorge along the main footpath in August 2003. Shown here is Jeremy Thompson, a technical adviser to the Lower Kihansi Environmental Management Project

Fig. 9.2 Concrete footbaths were installed in 2016 at the entrance of all of the spray wetlands and at the top and bottom of the Kihansi Gorge along the main footpath. Shown here is Joshua Malago a veterinary pathologist at Sokoine University of Agriculture who was trained at the Bronx zoo in amphibian histopathology and who provided histopathological analysis on amphibians between 2010 and 2019 for the Lower Kihansi Environmental Management and the Kihansi Catchment Conservation and Management Projects

with signs that were erected adjacent to the footbaths with instructions how to use the footbaths.

Following the joint LKEMP-World Bank supervisory wrap-up meeting on 12 August 2003, it was decided that a follow-up visit should be immediately conducted to the Kihansi Gorge to continue to assess the Kihansi Spray Toad populations in the spray wetlands and to collect skin swabs and soil and water samples.

A team comprising of biologists from the University of Dar es Salaam, representatives from the Ministry of Natural Resources, Jeremey Thompson, and I revisited all of the spray wetlands with the exception of the Mhalala Spray Wetland between the 14 and 15 August 2003. During this visit, numbers of Kihansi Spray Toads were recorded along line transects within the spray wetland vegetation, and total counts were conducted of toads on rock faces during both diurnal and nocturnal surveys. Water and soil samples were also collected from these wetlands and from the Kihansi reservoir and adjacent fields.

During these surveys, five skin swab samples were collected from Kihansi Spray Toads, three in the Upper Spray Wetland, and two in Mid-Gorge Spray Wetland. These samples were sent to Pisces Molecular a commercial laboratory in Boulder, Colorado, that specializes in applying molecular genetic techniques in wildlife disease diagnosis. The soil and water samples were sent to the government Chemist Laboratory in Tanzania.

Following the mid-August 2003 survey, the spray irrigation maintenance team continued to record through May 2004 numbers of Kihansi Spray Toads that were observed in the spray wetlands.

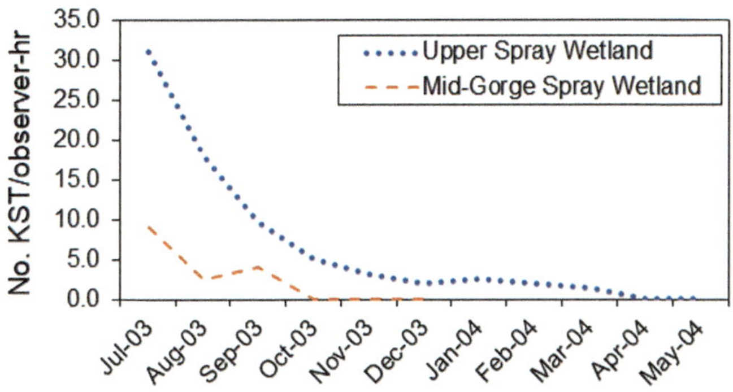

Fig. 9.3 Change in numbers of Kihansi Spray Toads per observer-hour in the Upper and Mid-Gorge Spray Wetlands between June 2003 and May 2004. (After CBSG 2008)

9.2 Post-June 2003 Decline of the Kihansi Spray Toad

The rapid decline of the Kihansi Spray Toad in the Kihansi Gorge that began during the second week in June 2003 was best documented in the Upper and Mid-Gorge Spray Wetlands (Fig. 9.3). These two spray wetlands had contained between October 2001 and June 2003 89% of the total population of the Kihansi Spray Toad in the Kihansi Gorge (Fig. 8.4).

The last recorded sightings for the Kihansi Spray Toad in the Kihansi Gorge included a single observation in May 2005 and two sightings in May 2006 (Mtui 2006), although no photographs nor voucher samples were taken of these last three sightings (Krajick 2006; Weldon and Mtui 2006). In 2009, IUCN formally listed the Kihansi Spray Toad as extinct-in-the-wild.

The dramatic and precipitous decline in numbers of Kihansi Spray Toad in the Kihansi Gorge was shocking particularly given that during the first week in June 2003 the estimated population of the Kihansi Spray Toad within the Kihansi Gorge was approximately 21,000 animals.

9.3 Proposed Hypotheses

Four prominent hypotheses have been proposed for the post-June 2003 collapse of the Kihansi Spray Toad population in the Kihansi Gorge. These hypotheses are disease, habitat alteration, a toxic substance, or an interaction between disease and a toxic substance. It had also been proposed, although not seriously considered, that seasonal aestivation and/or malicious human behavior may have contribute to the sudden decline of the Kihansi Spray Toad (Thompson 2003). However, these latter hypotheses were excluded due to the aseasonal environment of the spray wetlands and the difficulty for outside visitors to enter the gorge and spray wetlands undetected due to near daily presence of the sprinkler maintenance crew in the Kihansi Gorge.

In early September 2003, Pisces Molecular Laboratory in Boulder, Colorado, reported that skin swabs collected in August 2003 tested positive for *Batrachochytrium dendrobatidis* based on polymerase chain reaction (PCR) analysis from the two samples collected in the Mid-Gorge Spray Wetland but were negative for the three samples collected in the Upper Spray Wetland.

The Government Chemist Laboratory was unable to conduct an analysis for pesticides in the soil and water samples that had been collected in August 2003 in the Kihansi Gorge because the laboratory did not have the equipment required to detect pesticides in low concentrations. However, interviews conducted with farmers in the Kihansi catchment in mid-September 2003 did confirm that farmers use a variety of pesticides including endosulfan, a pesticide that is known to be highly toxic to aquatic organisms, to control stem borer disease in maize.[1]

9.3.1 Chytrid Fungus *Batrachochytrium dendrobatidis*

In November 2003, LKEMP contacted Ché Weldon (Fig. 9.4), a biologist from North-west University in South Africa with expertise in chytridiomycosis in African anurans, and requested his assistance in accessing the prevalence of chytridiomycosis in the Kihansi Gorge and surrounding area. In late November 2003, Ché Weldon,

Fig. 9.4 Ché Weldon, a biologist from North-West University in South Africa with expertise in chytridiomycosis in African anurans, surveyed amphibian species between 2003 and 2006 in the Kihansi Gorge for chytridiomycosis, trained Tanzanian scientists in the detection of chytridiomycosis, and assisted in the design and implementation of the disease transmission experiments

[1] In 2011, endosulfan was globally banned under the Stockholm Convention because of its toxicity to aquatic organisms although certain uses of the pesticide were permitted until 2017 (Hogue 2011).

Jeremy Thompson, and James Gibbs, who was in-country to assist LKEMP in designing an ecological monitoring program for the Kihansi Gorge, collected 45 tissue samples from five amphibian species over 2 days from four locations in the Kihansi Gorge as well as from one location in the Udagaji Gorge, located 2.5 kilometers to the east of the Kihansi Gorge. During the November 2003 survey, Ché Weldon and colleagues also visited the Mhalala Spray Wetland to assess from a distance the status of the Kihansi Spray Toad in this spray wetland. During this brief visit they were unable to detect any Kihansi Spray Toads in the Mhalala Spray Wetland.

In a follow-up laboratory analysis, Weldon detected chytridiomycosis in two of the five anuran species he collected in the Kihansi and Udagaji Gorges (Table 9.1).

In May 2006, Ché Weldon conducted a follow-up survey to assess again the prevalence of chytridiomycosis in anurans in the Kihansi and Udagaji Gorges. Over a 3-day period, he and Devolent Mtui, the LKEMP research and monitoring program officer, revisited the Kihansi and Udagaji Gorges. They collected 28 specimens from four amphibian species from eight locations in the Kihansi Gorge and from one location in the Udagaji Gorge. Chytridiomycosis was detected in *Anthroleptides yakusini*, in two locations in the Kihansi Gorge, the Upper Spray Wetland and Handaki Stream, as well as at the single sample site in the Udagaji Gorge (Table 9.2) (Weldon and Mtui 2006). Given that chytridiomycosis was detected in *A. yakusini* in both juvenile and adult animals in 2003 and 2006 and that *A. yakusini* did not display any adverse effects nor did the abundance of *A. yakusini* change between the 2003 and 2006 surveys, Ché Weldon concluded that *A. yakusini* was very likely a reservoir host of chytrid fungus in the Kihansi and Udagaji Gorges.

In 2020, Ché Weldon and colleagues conducted a retrospective analysis of 107 specimens of the Kihansi Spray Toad that were housed in the British Natural History Museum and in herpetological collections at the University of Dar es Salaam and at the University of the Western Cape (Weldon et al. 2020). In this analysis, they tested for the presence of *B. dendrobatidis* from skin brushing from paratypes (n = 18) at the British Museum and conducted a histological examination for the presence of *B. dendrobatidis* from toe clips from specimens housed at the University of Dar es Salaam and the University of the Western Cape (n = 89). He and colleagues also conducted a smaller retrospective analysis of eight *Nectophrynoides tornieri* which

Table 9.1 Distribution and prevalence of chytridiomycosis in anurans from Kihansi and Udagaji Gorges in November 2003

Location	Species	No. specimens examined	Prevalence (%)
Kihansi Gorge	*Arthroleptides* sp.	9	0
	Afrixalus fornasini	1	0
	Nectophrynoides tornieri	4	0
	Ptychadena anchietae	1	100
	Schoutedenella xenodactyla	17	0
Udagaji Gorge	*Arthroleptides* sp.	11	27.3
	Nectophrynoides tornieri	4	0

After Weldon (2004)

Table 9.2 Distribution and prevalence of chytridiomycosis in anurans in the Kihansi and Udagaji Gorges in May 2006 (After Weldon and Mtui 2006)

Location	Species	No. specimens examined	Prevalence (%)
Kihansi Gorge	*Arthroleptides yakusini*	11	18.2
	Arthroleptides stenodactylus	1	0
	Hyperolius puncticulatus	1	0
	Schoutedenella xenodactyla	8	0
Udagaji Gorge	*Arthroleptides yakusini*	7	28.6

After Weldon and Mtui (2006)

are found in vegetation immediately adjacent to the spray wetlands in the Kihansi Gorge. Their results were striking. All 99 *N. asperginis* (Kihansi Spray Toad) and 8 *N. tornieri* collected between 1986 and 2002 tested negative for *Bd*. On the other hand, five of the eight specimens of *N. asperginis* collected between August and November 2003 tested positive for chytrid fungus. These results combined with his earlier findings about the prevalence of *B. dendrobatidis* in other anuran species in the Kihansi and Udagaji Gorges (Tables 9.1 and 9.2), as well as James Gibbs' negative experimental results on the toxicity of the observed concentrations of endosulfan in reservoir sediments in Kihansi reservoir on the Kihansi Spray Toad (see below), provide compelling evidence that chytrid fungus was very likely the proximate cause for the sudden collapse and extinction-in-the-wild of the Kihansi Spray Toad.

Additional evidence for the Kihansi Spray Toad's lack of immunity to chytrid fungus comes from outbreaks of chytrid fungus in captive breeding facilities in the United States and Tanzania. The first recorded outbreak occurred in April 2007 in the Bronx zoo and resulted in the death of 49 of 49 animals in one cage (CBSG 2008).

The second outbreak of chytrid fungus occurred in November 2012 in the Kihansi captive breeding facility, located at the base of the Kihansi Gorge, and resulted in the death of 351 animals in 17 cages or 62% of the captive population (Makange et al. 2014). In this outbreak all of the dead animals displayed signs of chytridiomycosis, and of the 49 dead animals that were collected, all individuals tested positive for chytrid fungus based on PCR analysis (Makange et al. 2014).

The third outbreak of chytridiomycosis occurred in the University of Dar es Salaam captive breeding center in January 2013 and resulted in the death of 23 individuals in one cage, and the fourth outbreak occurred in April 2016 in the Kihansi captive breeding center and resulted in the death of 40 individuals. Gerald Misinzo, a microbiologist at Sokoine University of Agriculture, tested specimens for chytrid fungus from the latter two outbreaks using PCR analysis, and the results were positive in both cases.

The sudden collapse of the Kihansi Spray Toad from approximately 18,000 in the Upper Spray Wetland to fewer than 50 individuals over a 2-month period between June and August 2003 was also consistent with the hypothesis that *B. dendrobatidis* was a novel disease to the Kihansi Spray Toad rather than an endemic pathogen that became more virulent due to climate change (Fisher and Garner 2020; Weldon et al. 2020). The extinction-in-the-wild of the Kihansi Spray Toad was the first recorded extinction of an amphibian species in Africa due to chytridiomycosis (Fisher and Garner 2020).

Recent sequencing of *B. dendrobatidis* from two archived Kihansi Spray Toads that were collected in 2004 during the population crash revealed that both individuals were infected with the *Bd*CAPE strain rather than the more widespread *Bd*GBL strain (Sewell et al. 2024). Furthermore, the sequenced genotype from these two animals was typical of *Bd*CAPE strains from Southern Africa that are known to be virulent. This same analysis also suggested, based on an estimated time to the most common ancestor (tMTRC) analysis, that the introduction of the *Bd*CAPE strain into the Kihansi Gorge occurred sometime after the emergence of the virulent strain in Africa. The estimated mean emergence date of the more virulent *Bd*CAPE strain in Africa was 1991 (range = 1944–2003) (Sewell et al. 2024).

The extinction-in-the wild of the Kihansi Spray Toad is also the first extinction of an amphibian species that has been attributed to the *Bd*CAPE strain, although the *Bd*CAPE strain was previously identified to be responsible for the decline of the Mallorcan Midwife Toad *Alytes muletensis* following its accidental introduction into the reintroduced Balearic population (Sewell et al. 2024).

Finally, this analysis examined the geographic distribution of *Bd* and the *Bd*CAPE strain in the Udzungwa Mountains based on two surveys conducted between 2004–2005 and 2013–2015 within 30 km of the Kihansi Gorge. In the first survey, *Bd* was detected in 35 of the 554 (6.3%) specimens and 10 of 48 (8.7%) species. In the second survey, 15 of the 60 (25%) swabs tested positive for *Bd* all of which were the *Bd*CAPE strain (Sewell et al. 2024). However, in this Udzungwa-wide survey, none of the amphibian species that tested positive for *B. dendrobatidis*, including *Nectophrynoides tornieri* a closely related species, displayed signs of chytridiomycosis or have experienced significant population declines. Thus Sewell and colleagues hypothesize that the population crash of the Kihansi Spray Toad in June 2003 was likely a result of an interaction between chytridiomycosis and some form of environmental perturbation.

9.3.2 Habitat Alteration

A second hypotheses that was proposed for the sudden collapse of the Kihansi Spray Toad population in the Kihansi Gorge in June 2003 was habitat alteration or an interaction between chytrid fungus and habitat alteration (Weldon et al. 2020). However, there is countervailing evidence for this hypothesis. Firstly, the Kihansi Spray Toad population in the Upper Spray Wetland rebounded in early June 2003 to pre-diversion population levels in this wetland following the installation of the spray irrigation systems (Sect. 8.3), suggesting that habitat alteration was not limiting the population recovery of the Kihansi Spray Toad, at least in the Upper Spray Wetland. And secondly and more importantly, the spray wetland vegetation in the Mhalala Spray Wetland had not been altered by a reduction in flow of the Kihansi River following the commissioning of the Lower Kihansi hydropower facility. The Mhalala Spray Wetland was maintained by spray generated by a waterfall along the Mhalala stream, a tributary of the Kihansi River. However, the Kihansi Spray Toad population in Mhalala Spray Wetland also crashed in 2003, sometime between June and November 2003.

9.3.3 Toxic Contaminant

A third prominent hypothesis for the sudden collapse of the Kihansi Spray Toad population in the Kihansi Gorge was a toxic contaminant.

One of the objectives of the early June 2003 survey of the Kihansi Spray Toad population in the Upper Spray Wetland was to assess the behavioral response of the Kihansi Spray Toad to intermittent bypass flow releases of 12 m^3/s and 16 m^3/s as well as to document the spatial dispersion of the spray generated by these bypass releases. Intermittent bypass flow releases had been proposed as a possible mitigation option to maintain the spray wetland vegetation in case the spray irrigation systems were to fail.

Under the IREM project, intermittent bypass flow short-term releases of 4 m^3/s, 6 m^3/s, and 8 m^3/s had revealed that intermittent releases of these magnitudes were insufficient to generate sufficient spray to cover the Upper Spray Wetland. Thus, LKEMP and the World Bank had supported NORPLAN's proposal to assess whether intermittent bypass flow releases of 12 m^3/s and 16 m^3/s could generate sufficient spray to cover the Upper Spray Wetland. Although releases of 12 m^3/s and 16 m^3/s did provide sufficient spray to cover the Upper Spray Wetland, it was concluded after June 2003 that intermittent bypass flow releases were not a viable option for maintaining the spray wetlands for several reasons: firstly, the high likelihood of regularly releasing toxicants (see below); secondly, intermittent bypass flow releases would be insufficient in terms of duration to maintain the spray wetlands; and lastly intermittent bypass flow releases would greatly reduce the electrical generating capacity of the Lower Kihansi hydropower facility.

In order to conduct the intermittent bypass flow releases of 12 m^3/s and 16 m^3/s at the end of May and early June 2003, the hydroelectric facility operators were required to open a flushing gate on the Kihansi Dam. The flushing gate was designed to permit the flushing of soil sediments that accumulate over time behind the dam.

During the high-flow intermittent releases that were conducted over a 4-hour period on 31 May and 1 June 2003, observers noted substantial quantities of sediment in the released water that were of a sufficient magnitude to darken the color of the Kihansi River for several hours following the high-flow bypass release.

Following the June 2003 crash of the Kihansi Spray Toad population in the Kihansi Gorge, farmers in two villages located upstream from the dam were interviewed about their use of pesticides. Farmers in these two villages reported using up to 5 kg of pesticide per ha to control insect pests on maize, cabbage, and beans. One of the pesticides that farmers in the catchment reported using was endosulfan, a pesticide that binds to soil sediments and is known to be extremely toxic to fish.

Because of the difficulty that the Government Chemist Laboratory had in detecting pesticides at low concentrations, LKEMP requested in September 2004 that the Tropical Pest and Research Institute (TPRI) based in Arusha, Tanzania, collect and analyze new samples.

Between June and September 2005, TPRI collected water, soil, and fish samples along an elevational gradient that began above the dam, including the reservoir, and extended through the Upper, Lower, and Mid-Gorge Spray wetlands. TPRI also re-interviewed farmers about pesticide use and confirmed that pesticides were widely used in the Kihansi catchment to control stem borer disease in maize.

The collected soil and water samples were distributed among three laboratories—TPRI, South African Bureau of Standards (SABS), and the Netherlands Organization for Applied Scientific Research (TNO). At each of these laboratories, gas chromatography coupled with mass spectrometry was used to identify pesticides within the samples.

Results from these analyses indicated higher concentrations of pesticides in sediment, soil, and fish samples than in water (Table 9.3). The concentration of endosulfan was highest in reservoir sediments (10–26 μg/kg dry weight). DDT metabolites and other organochlorines including α-lindane and Aldrin were also detected in trace quantities in soil, water, and fish samples collected above, within, and below the reservoir (TPRI 2005). In 2009, TPRI repeated the analysis within and below the dam including in the Mhalala tributary. In this 2009 survey, samples were equally divided among TPRI and the Netherlands Organization for Applied Scientific Research.

In the 2009 analysis, an even larger number ($n \approx 160$) of pesticides and metabolites residues including endosulfan, DDT metabolites, and other organophosphates and organochlorines were detected in water, soil, and fish samples indicating a continuing if not an expanded use of pesticides in the Kihansi River catchment (TPRI 2009). As in the 2005 survey, soil, reservoir sediment, and fish samples collected in 2009 had again higher concentrations than water samples.

While many of the DDT metabolites are known to be endocrine disruptors to amphibians (Kloas 2002; Hoffmann and Kloas 2016), what was of particular and continuing concern was the comparatively high concentrations of endosulfan (5–50 μg/kg dry weight) that were detected again in the 2009 survey in soil, reservoir sediment, and water samples. This was of particular concern because by 2009 plans were being developed to reintroduce the Kihansi Spray Toad back to the spray wetlands.

In 2010, LKEMP commissioned James Gibbs and two of his graduate students at SUNY, Syracuse, to assess the toxicity of endosulfan to the Kihansi Spray Toad. James Gibbs and his graduate students reported that signs of toxicity to endosulfan in the Kihansi Spray Toad were only observed in concentrations exceeding those

Table 9.3 Pesticide residue levels detected in sediment and soil samples collected in the Kihansi River, reservoir, and at six locations in the Kihansi Gorge between August and September 2005

	Sediment μg/kg dry weight				Soil μg/kg dry weight	
Pesticide Residues	Reservoir	Mid- Gorge	Upper Gorge	Upstream of dam	Upper Gorge	Upstream of dam
Hexachloro-1,3-butadiene	<0.5	<0.5	<0.5	<0.5	<0.5	<0.5
Pentachlorobenzene	<0.5	<0.5	<0.5	<0.5	<0.5	<0.5
α-lindane	<0.5–5.6	<0.5–7.6	<4.3	<0.5–5.8	<0.5–9.1	<0.5–5.8
Hexachlorobenzene	<0.5	<0.5	<0.5	<0.5	<0.5	<0.5
β-Lindane	<2	<2	<2	<2	<2	<2
γ-Lindane	<2	<2	<2	<2	<2	<2
δ-Lindane	<2	<2	<2	<2	<2	<2
Heptachlor	<0.5	<0.5	<0.5	<0.5	<0.5	<0.5
Aldrin	<0.5–1	<0.5–4.6	<0.5–9	<0.5	<0.5–0.9	<0.5
Heptachlor epoxide	<0.5	<0.5	<0.5	<0.5	<0.5	<0.5
o,p'-DDE	<0.5	<0.5	<0.5	<0.5	<0.5	<0.5
Endosulfan I	**<10–26**	**<10**	**<10**	**<10**	**<10**	**<10**
p,p'-DDE	<0.5–12	<0.5–0.6	<0.5	<0.5	<0.5	<0.5
Dieldrin	<2	<2	<2	<2	<2	<2
o,p'-DDD	<0.5	<0.5	<0.5	<0.5	<0.5	<0.5
Endrin	<10	<10	<10	<10	<10	<10
p,p'-DDD	<0.5 0.6	<0.5	<0.5	<0.5	<0.5	<0.5
o,p'-DDT	<0.5	<0.5	<0.5	<0.5	<0.5	<0.5
p,p'-DDT	<0.5	<0.5	<0.5	<0.5	<0.5	<0.5

Concentrations of endosulfan a pesticide known to be extremely toxic to aquatic organisms are highlighted in bold. After TPRI (2005)

that TPRI had been detected in soil, water, and reservoir sediments in the Kihansi Gorge (Gibbs et al. 2011). They estimated that the 4-day median lethal concentration (LC50) of endosulfan for the Kihansi Spray Toad was 360 ppb (Gibbs et al. 2011).

These results thus suggest that the proximate cause of the sudden collapse of the Kihansi Spray Toad population in the Kihansi Gorge in 2003 was either *B. dendrobatidis* or possibly an interaction between *B. dendrobatidis* and exposure to a toxic contaminant. However, an interaction between *B. dendrobatidis* and a toxic contaminant has not been tested. On the other hand, there is countervailing evidence to an interaction between *B. dendrobatidis* and a toxic contaminant, and this comes again from the Mhalala Spray Wetland. The Mhalala Spray Wetland is located 180 m upslope from the Kihansi River along the Mhalala stream, a tributary of the Kihansi River. And thus, unless there were strong winds blowing through the gorge in the direction of the Mhalala Spray Wetland during the high flow intermittent bypass release study, the Mhalala Spray Wetland would have very likely been located too far from the Kihansi River to have received any pesticide-contaminated spray. However, the Kihansi Spray Toad population in the Mhalala Spray Wetland also collapsed in mid- to late 2003.

Additional countervailing evidence to an interaction between *B. dendrobatidis* and a toxic contaminant comes from the four outbreaks of chytridiomycosis in captive Kihansi Spray Toad populations in the US and Tanzanian captive breeding facilities. In all of these outbreaks, deaths of animals occurred in the absence of a toxic contaminant.

Thus, I find the most compelling hypothesis for the population collapse and extinction-in- the-wild of the Kihansi Spray Toad was *Batrachochytrium dendrobatidis* and that *Bd* was a novel disease (Weldon et al. 2020). The extremely small historic geographic range (<0.5 ha) of the Kihansi Spray Toad within a remote and an uninhabited valley could have very likely provided historic protection for the Kihansi Spray Toad from *Bd*. This also could explain the Kihansi Spray Toad's apparent unique vulnerability among amphibian species in the Udzungwa Mountains to chytridiomycosis.

9.4 External Review Panel, 2004

Following the collapse of the Kihansi Spray Toad population in June 2003, the government of Tanzania and LKEMP requested a reconvening of the external review panel to seek their input and recommendations. In January 2004, a 2-day meeting was organized comprising of the same panelists that met in January 2003 as well as Ché Weldon from the Northwest University in South Africa, Sam Lee the Kihansi Spray Toad attendant at the Bronx zoo, and John Gerstle a NORPLAN consultant. The Tanzanian delegation included Raphael Mollel, Permanent Secretary in the Vice President's Office; Magnus Ngoile, Director General of the National Environment Management Council; Wilfred Sarunday LKEMP project coordinator; and Jeremy Thompson technical adviser to LKEMP. The World Bank team was represented by Ladisy Chengula, task team leader for LKEMP, and myself, a consultant to the Word Bank.

The External Review Panel proposed that the objectives for ecological monitoring and research in the Kihansi Gorge remain the same as had been proposed in January 2003. They also made and prioritized recommendations for future research (Table 9.4). Lastly, the External.

Review Panel made several additional policy recommendations in response to the population crash of the Kihansi Spray Toad in June 2003. These were the following:

1. The pathogen hygiene protocols outlined in Weldon (2004) should be implemented immediately in order to reduce the likelihood of transmission of disease between wetlands inside of the Kihansi Gorge and potentially to other areas outside of the gorge.
2. Because of the tremendous risk that chytrid fungus poses to amphibian populations worldwide, the scientific community should be informed about the documented presence of chytrid fungus in the Kihansi Gorge.
3. Given recent dramatic declines in the wild and captive populations of the Kihansi Spray Toad, there is now a much higher likelihood of species extinction.

Table 9.4 Taxon-specific recommendations for monitoring and research proposed by the External Review Panel and workshop participants in January 2004

Taxa	Recommendation for future research
Kihansi Spray Toad	1. Determine whether PVC pipe placed in permanent positions can be used to more easily and powerfully sample Kihansi Spray Toad numbers in the wetland vegetation. The methodology has been used to successfully improve frog monitoring in other areas. Priority = 1.6
Insects	1. Develop a monitoring protocol for Diptera. Priority = 1.6
Spray wetland vegetation	1. Increase the density and/or change the distribution of sprinklers to better recreate natural precipitation and soil moisture levels. Experimentation with water jets, soaker hoses, or drip irrigation could also be used in conjunction with sprinklers and jets. Data could be collected on responses of soil characteristics, plants, insects, and toads. Priority = 1.0
Kihansi Spray Toad laboratory experiments	1. Nutritional content studies. Priority = 1.0
Chytrid fungus studies	1. Identify the genetic strain of *Batrachochytrium dendrobatidis* in the Kihansi Gorge. Priority = 1.6

The priority scale ranges from 1.0 to 3.0 with 1.0 being the highest priority and 3.0 being the lowest priority

Therefore, in order to conserve as much of the genetic diversity of the Kihansi Spray Toad in perpetuity, it is strongly recommended that efforts be undertaken immediately to extract cells from ill and low reproductive potential individuals in captivity to establish a live cell line. This offers the possibility of potentially recovering the species in the future.

4. To assist current and future management in the Kihansi Gorge, the ecological monitoring recommendations included in draft final report (Gibbs 2003) concerning the strongly recommended indicators, data collection, analysis, and reporting should be adopted immediately. LKEMP should identify an individual who will be responsible for managing and backing up the monitoring and research databases. As part of this database management, data need to be compiled and stored in a fashion that will allow easy and rapid access and analysis.
5. Mitigation measures should be initiated immediately in the Kihansi Gorge if a 50% drop is observed from the previous year and/or baseline for any attributes of indicators as specified in the Gibbs (2003) report.
6. No exotic species should be introduced into the Kihansi Gorge, reservoir, or catchment area for any purpose including biomonitoring.

9.5 Establishment of a Final Water Right

Since the commissioning of the Kihansi power station in December 1999, TANESCO had been operating the hydroelectric facility in the absence of a final water right. Additionally, it was also operating the hydroelectric facility in non-compliance with the provisional water right which stipulated the maintenance of a minimum bypass flow of 7.0 m^3/s. Consequently, the second major project component of LKEMP was focused on establishing through a transparent process TANESCO's final water right for the operation of the Kihansi power station and the required monitoring for compliance. As part of this process, LKEMP provided support for the monitoring of the interim bypass flow of 1.5–1.9 m^3/s, the continuing assessment of the value of various intermittent bypass flow releases to conserve the spray wetlands, and the expansion of the spray irrigation system in the three largest spray wetlands. This component also focused on developing a national policy framework and methodology for allocating water for the environment, i.e., environmental flows.

In June 2005, the Rufiji Basin Water Office issued a final water right to TANESCO. Among the eight stipulated conditions of the final water right, four were of particular relevance to the implementation of the species recovery program for the Kihansi Spray Toad. These conditions were that TANESCO (1) maintain a minimum bypass flow of 1.5–2.0 m^3/s in the Kihansi River immediately downstream of the dam; (2) keep daily records of water levels and storage in the reservoir, turbine discharges, bypass flows, and spills and provide this information to the Rufiji Basin Water Office on a weekly basis; (3) shall not carry out reservoir sediment flushing without approval from the Rufiji Basin Water Officer and the National Environmental Management Council; and (4) shall ensure that the water that is returned to the river is undiminished and unpolluted. The final water right also stipulated that a review and potential revision of the water right will be carried out at least once every 5 years.

To monitor compliance with the final water right, the Rufiji Basin Water Office installed automatic data loggers and staff gauges to monitor water flows and levels in the Kihansi River. The staff gauges were read twice a day, while the data loggers were downloaded at intervals of a week to a month (Arcadis Euroconsult 2004).

9.6 Updating the Environmental Management Plan

A legal covenant attached to the 1993 World Bank loan to the Tanzanian government for the Lower Kihansi Hydroelectric Project was the requirement to develop and implement an Environmental Management Plan to address any environmental and social impacts arising from the hydropower project. In 1993 it was assumed that most if not all of the environmental and social impacts would result from the impoundment of water behind the dam. However, by 1996 it was apparent that the dam would have much broader environmental impacts and particularly in the light of the discovery of the Kihansi Spray Toad and several other endemic plant species. As a result, an additional condition for the final water right for the Kihansi hydropower station was that the Environmental Management Plan would need to be updated.

Under the updated Environmental Management Plan, it was highlighted that TANESCO was responsible for financially supporting over the dam's life span, the sprinkler systems, bridges, staircases, walkways, and the captive breeding and reintroduction program for the Kihansi Spray Toad including the monitoring of key ecological indicators. The costs associated with the Kihansi Spray Toad recovery program represented <1% of the annual revenue generated from the Lower Kihansi hydropower facility (Arcadis Euroconsult 2004). Yet, TANESCO has to date restricted their financial support of the Kihansi Spray Toad recovery program to maintaining the sprinkler systems and gorge infrastructure.

References

Arcadis Euroconsult (2004) Lower Kihansi hydropower project updated environmental management plan. Dar es Salaam University Press, Dar es Salaam

Berger L, Speare R, Daszak P et al (1998) Chytridiomycosis causes amphibian mortality associated with population declines in the rain forests of Australia and Central America. PNAS 95(15):9031–9036

CBSG (2008) Population and habitat viability assessment for the Kihansi Spray Toad. Apple Valley

Fisher MC, Garner TWJ (2020) Chytrid fungi and global amphibian declines. Nat Rev Microbiol 18:332–343. https://doi.org/10.1038/s41579-020-0335-x

Gibbs JP (2003) A detailed ecological monitoring protocol for the Lower Kihansi Gorge ecosystem. Draft report to the Lower Kihansi Environmental Management Project

Gibbs JP, Reeve B, Radell C (2011) Evaluation of toxicity of Endosulfan for Kihansi Spray Toads: direct exposure effects and interactions with nutritional status and infection by pathogenic chytrid fungus. An addendum to the final report to the Lower Kihansi Environmental Management Project

Hoffmann F, Kloas W (2016) p,p′-Dichlorodiphenyldichloroethylene (p,p′-DDE) can elicit antiandrogenic and estrogenic modes of action in the amphibian Xenopus laevis. Physiol Behav 167:172–178. https://doi.org/10.1016/j.physbeh.2016.09.012

Hogue C (2011) Endosulfan banned worldwide. Chem Eng News 89:15

Kloas W (2002) Amphibians as a model for the study of endocrine disruptors. Int Rev Cytol 216:1–57. https://doi.org/10.1016/S0074-7696(02)16002-5

Krajick K (2006) The lost world of the Kihansi toad. Science 311:1230–1232. https://doi.org/10.1126/science.311.5765.1230

Makange M, Kulaya N, Biseko E et al (2014) Batrachochytrium dendrobatidis detected in Kihansi spray toads at a captive breeding facility (Kihansi, Tanzania). Dis Aquat Org 111:159–164. https://doi.org/10.3354/dao02775

Mtui D (2006) Monitoring Kihansi Gorge 2004–2006. Lower Kihansi Environmental Management Project

NORPLAN (2002) Lower Kihansi Hydropower Project. Immediate rescue and emergency measures. Final specialist report: Amphibian studies. Report produced for Tanzania Electric Supply Company Ltd. (TANESCO), Dar es Salaam, Tanzania

Sewell TR, Van DL, Ghosh PN et al (2024) Archival mitogenomes identify invasion by the Batrachochytrium dendrobatidis CAPE lineage caused an African amphibian extinction in the wild. Proc Biol Sci 291:20241157

Thompson J (2003) Fact sheet draft version 1. Lower Kihansi Environmental Management Project

TPRI (2005) Toxicological surveys for the presence of pesticides at Kihansi Gorge and its catchment areas. Lower Kihansi Environmental Management Project

TPRI (2009) Monitoring for the presence of pesticide residues at Kihansi Gorge and its catchment areas. Lower Kihansi Environmental Management Project

Weldon C (2004) Chytridiomycosis risk assessment in the Kihansi and Udagaji Gorges with special reference to the Kihansi Spray Toad. Draft final report for the Lower Kihansi Environmental Management Project

Weldon C, Mtui D (2006) Follow-up survey at Kihansi Gorge for the presence of the amphibian chytrid fungus. Lower Kihansi Environmental Management Project

Weldon C, Channing A, Misinzo G, Cunningham AA (2020) Disease driven extinction in the wild of the Kihansi spray toad, Nectophrynoides asperginis. African J Herpetol 69:151–164. https://doi.org/10.1080/21564574.2020.1752313

Captive Breeding of the Kihansi Spray Toad in US Zoos

10

Abstract

In November 2000, Wildlife Conservation Society personnel transferred 499 Kihansi Spray Toads from the Kihansi Gorge to the Bronx and Detroit zoos, thus establishing the captive population of the Kihansi Spray Toad in US zoos. Although the founding population in the US zoos declined initially due to a lungworm, this was successfully treated, and by July 2002 the captive population of the Kihansi Spray Toad was 580. However due to limited space and resources to manage the growing population, a decision was made to reduce the size of the captive population in the Bronx and Detroit zoos by transferring animals to other zoos and by altering the misting schedule within the cages. Following the collapse of the Kihansi Spray Toad population in the Kihansi Gorge in June 2003, there was an urgency to build up the captive population in the US zoos. Between 2002 and early 2004, the captive population in US zoos continued to decline, and by March 2004 the captive population was 71. To prevent a possible global extinction of the Kihansi Spray Toad, an attempt was made to establish a cell line, which failed. However, as a result of a refinement of the husbandry protocols for the Kihansi Spray Toad, the captive population after March 2004 slowly began to recover, and by late 2007 the captive population was rapidly expanding. By 2010 the captive population of the Kihansi Spray Toad in the Toledo and Bronx zoo was >6400, one of the largest captive populations for any extinct-in-the-wild species worldwide.

During the last week of November 2000, personnel from the Wildlife Conservation Society collected and transferred 499 adult Kihansi Spray Toads from the Kihansi Gorge to New York. Most of the animals that were collected were from the Upper

W. D. Newmark, *The Kihansi Spray Toad* (Nectophrynoides asperginis): *Crisis Biology in the 21st Century*, Ecological Studies 254, https://doi.org/10.1007/978-3-031-89599-9_10

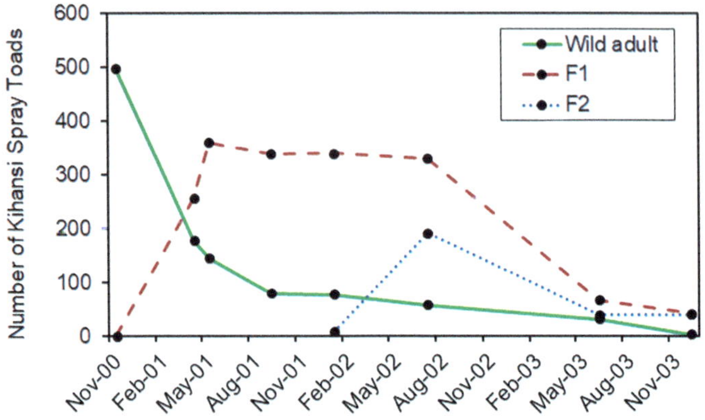

Fig. 10.1 Change in numbers of wild adult, F1, and F2 generations of the Kihansi Spray Toad between November 2000 and November 2003 in US captive breeding facilities. (Source: Lee 2002; Lee et al. 2006)

Spray Wetland with the remainder being collected close to Jabali intake—the intake for the Upper Spray Wetland gravity-fed spray irrigation system.

After the arrival of the Kihansi Spray Toads in New York, 269 toads were retained at the Bronx Zoo, and 230 were transferred to the Detroit zoo's National Amphibian Conservation Center. Over the following 6 months, the wild founder population declined in both of the captive breeding facilities. In the Bronx zoo, the wild adult population declined from 269 individuals to 48 and in the Detroit zoo declined from 203 individuals to 76 (Lee 2002; Lee et al. 2006).

Necropsies conducted in-house and at Cornell University attributed the majority of the deaths to a lungworm caused by a Rhabditiform nematode which had been present in the animals collected in the Kihansi Gorge (Lee et al. 2006). Adult and juvenile animals were treated with three different drugs of which fenbendazole proved to be most effective (Lee et al. 2006). Food items were also dusted with pulverized fenbendazole. Fortunately, none of the F1 generation, which were born within 1 month of arrival in the United States (Fig. 10.1), were infected by the lungworm. By July 2002, the captive population in US zoos had rebounded to 580 animals comprising of 58 wild adults, 330 F1, and 192 F2 (Figs. 10.1 and 10.2).

10.1 Decline in the Captive Breeding Population, July 2002–March 2003

Beginning in July 2002, the captive Kihansi Spray Toad population in the US zoos began to decline again (Fig. 10.2). The initial decline over this time period was a result of a decision by the Bronx and Detroit zoos to reduce the size of their captive Kihansi Spray Toad populations because of limited space, staff, and financial resources to manage the rapidly growing captive population.

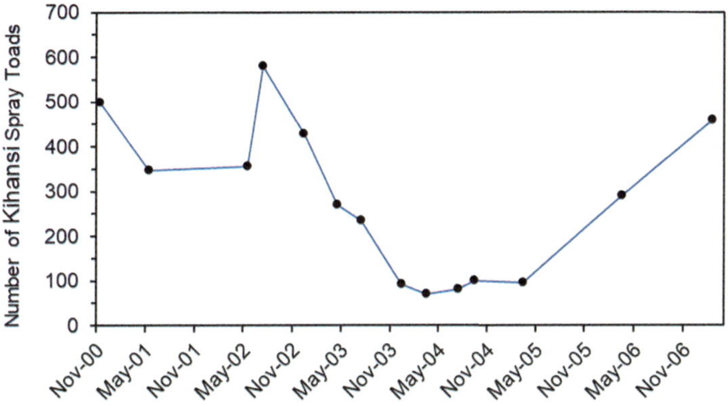

Fig. 10.2 Change in numbers of the Kihansi Spray Toad in US captive breeding facilities between November 2000 and December 2006. (Source: Lee 2002; Borek 2008)

In the early 2000s, amphibian captive breeding facilities worldwide were being forced to house an increasing number of amphibian species as a result of the dramatic decline of amphibians in the wild due to habitat loss, environmental contaminants, and emerging diseases (Berger et al. 1998; Stuart et al. 2004). The Kihansi Spray Toads that were housed in the Bronx and Detroit zoos were competing for space and resources with other endangered captive amphibian species that were being brought into these captive breeding facilities.

In early 2001, the Bronx and Detroit zoos sent out a request to a number of other zoos in the United States about their interest and willingness to house a captive population of the Kihansi Spray Toad. In May 2001, the Bronx zoo transferred 24 animals to the Oklahoma City zoo, 38 animals in July 2001 to the Baltimore zoo, and 20 animals in September 2003 to the Buffalo zoo (Table 10.1). The Detroit zoo also initiated a transfer of animals from their facility. In February 2002, the Detroit zoo sent 30 animals to the Toledo zoo which initiated the Toledo zoo's long-term involvement in managing a captive population of the Kihansi Spray Toad.

In a further attempt to reduce the population growth rate of the Kihansi Spray Toad in the Bronx zoo, because of limited resources and because the wild population of the Kihansi Spray Toad in the Kihansi Gorge was rebounding by mid-2002 (Fig. 8.2), the misting schedule of the caged Kihansi Spray Toad population in the Bronx zoo was reduced between July and September 2002 from 240 minutes per day to 24 minutes per day. This alteration in the misting schedule was successful in reducing the number of births from 171 over a 20-day period in June and July 2002 to 23 over a 153-day period between September 2002 and February 2003 (Lee et al. 2006).

However, everything changed in June 2003 with the sudden collapse over the following 2–3 months of the Kihansi Spray Toad population in the Kihansi Gorge. Rebuilding the captive breeding population of the Kihansi Spray Toad in the US captive breeding facilities immediately became the highest priority for saving the Kihansi Spray Toad. Although establishing a captive population of the Kihansi

Table 10.1 Summary of transfers of the Kihansi Spray Toad between US zoos between May 2001 and November 2005

Date	Source institution	Transfer represented source institution's remaining population	Receiving institution	Number of animals transferred
May 2001	Bronx	No	Oklahoma City	43
August 2001	Bronx	No	Baltimore	38
February 2002	Detroit	No	Toledo	30
September 2003	Bronx	No	Buffalo	20
February 2004	Detroit	Yes	Toledo	4
March 2004	Baltimore	Yes	Bronx	15
March 2004	Buffalo	Yes	Bronx	15
November 2005	Bronx	No	Toledo	25

Source: Lee (2003), Borek (2008)

Spray Toad in 2000 had been controversial in Tanzania, the importance and value of having an assurance population became immediately apparent to all stakeholders.

10.2 Consolidation of the Captive Populations

It was initially thought that by reinstating the former misting schedule in the cages housing the Kihansi Spray Toad, the captive population would quickly rebound. Between June and September 2003, the Kihansi Spray Toad attendants in the Bronx zoo increased the misting schedule from 52 minutes per day to 300 minutes per day. Yet over this time period, the number of births remained effectively unchanged ($n = 25$) from that recorded under a misting schedule of 52 minutes per day (Lee et al. 2006).

A decision was made in early 2004 by the US zoos holding the captive Kihansi Spray Toad populations to consolidate their captive populations into two zoos, the Bronx and Toledo. Between February and March 2004, the Detroit zoo transferred their remaining animals ($n = 4$) to the Toledo zoo. In March 2004, the Baltimore and Buffalo zoos each returned 15 animals to the Bronx zoo (Table 10.1). The Oklahoma City zoo captive population that had been established in May 2001 by animals from the Bronx zoo had collapsed by December 2003.

However, the consolidation of the captive populations of the Kihansi Spray Toad into two zoos did not result in an immediate increase in the captive population size, and by March 2004 the total captive population of the Kihansi Spray Toad in the Bronx and Toledo zoos had declined to 71 individuals.

10.3 Efforts to Establish a Cell Line

Given the dire situation in early 2004 of the Kihansi Spray Toad both in captivity and in the wild, the Bronx zoo proposed to establish a cell line from the captive Kihansi Spray Toad population in their zoo to conserve genetic material in case the Kihansi Spray Toad were to go extinct. The proposal was endorsed by the Tanzanian government and the External Review Panel, which recommended that a cell line be established as soon as possible. Following the receipt of permission from the Tanzanian Wildlife Division, the Bronx zoo collected tissues from six animals that were sent in early November 2004 to Coriell Cell Repositories in New Jersey. However, none of the cell lines persisted. Fortunately by mid-2005, the captive Kihansi Spray Toad population in the Bronx and Toledo zoos began to recover.

10.4 Post-2004 Recovery of the Captive Breeding Populations

In March 2004, the combined captive population of the Kihansi Spray Toad in the Bronx and Toledo zoos had declined to 71 individuals, and an attempt to establish a cell line in late 2004 had failed. In a valiant effort to increase the captive population, the spray toad attendants at the Toledo and Bronx zoos refined key husbandry parameters which included (1) switching from UV black lights to full-spectrum halogen lights; (2) altering the vitamin and mineral supplements; (3) changing the substrate in plastic cages from gravel to moss; and (4) increasing the misting schedule from 300 minutes per day to 24 hours per day (Fig. 10.4) (Lee et al. 2006; Herman 2010). Because of the small size of the captive population, the spray toad attendants were forced to combine generations within a cage, and thus after 2004 individual generations were no longer identifiable in the two zoos (Borek 2008).

In response to these changes in husbandry, the captive Kihansi Spray Toad population in the Bronx and Toledo zoos began to increase, although initially slowly (Fig. 10.3). By February 2007 the captive population of the Kihansi Spray Toad in the two zoos had reached 460 animals which was the approximate number of animals that had been originally collected in the wild in late 2000. By late 2007, the Kihansi Spray Toad population in the two captive breeding facilities began to rapidly increase (Fig. 10.4) and particularly in the Toledo zoo. By June 2010 the combined captive breeding population of the Kihansi Spray Toad in the Bronx and Toledo zoo was >6400 (Fig. 10.3) which represented one of the largest captive populations of any extinct-in-the-wild species worldwide.

During this period of rapid growth, one of the more pronounced differences that emerged between the captive and historic wild population was the timing of births. Long-term census data of the Kihansi Spray Toad at the Toledo Zoo revealed that

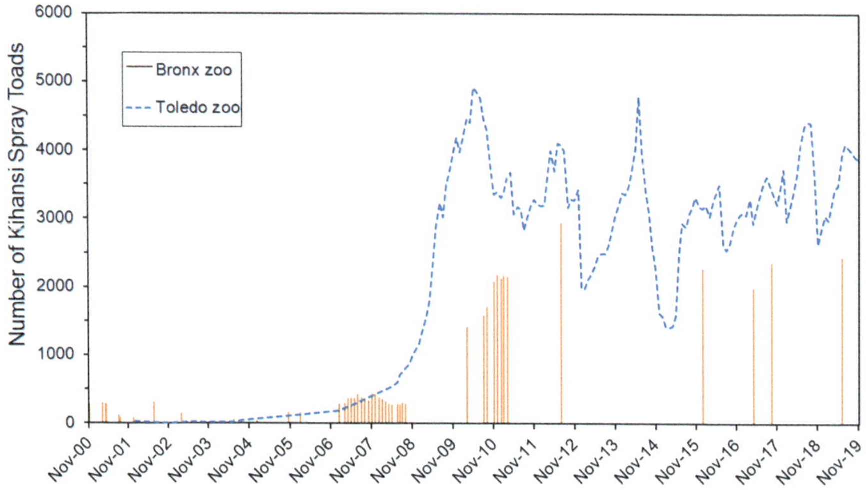

Fig. 10.3 Change in the captive populations of the Kihansi Spray Toad in the Bronx and Toledo zoos between November 2000 and December 2019. The Toledo zoo censused their captive population monthly, while the Bronx zoo intermittently censused their captive population. Declines in the Kihansi Spray Toad population after 2010 in the Toledo Zoo were largely a result of holding animals back from development and transfers of animals to Tanzania. (Source: Borek 2008; Bronx Zoo 2011, 2014, 2017, 2019; Odum 2018; Chastain 2019)

Fig. 10.4 Ladisy Chengula, World Bank task team leader of the Lower Kihansi Environmental Management Project between 2002 and 2007, inspecting a cage in the Bronx zoo housing the Kihansi Spray Toad

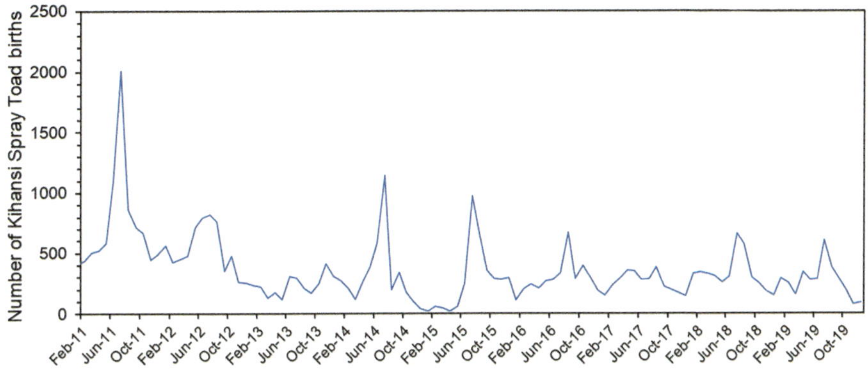

Fig. 10.5 Monthly births of the Kihansi Spray Toad between 2011 and 2019 in the Toledo captive breeding facility. (Source: Odum 2018, Chastain 2019)

births in their facility were highly seasonal with peak number of births occurring between June and August (Odum 2018) with a mean (± SD) interval between peak births of 365 (±73) days (Fig. 10.5). In the wild the number of births of the Kihansi Spray Toad peaked between January and February (Channing et al. 2006).

Indeed, the population of the Kihansi Spray Toad in the Toledo zoo was growing so rapidly between 2007 and 2010 that Andy Odum, curator of herpetology at the Toledo zoo, was concerned about the risk of a population collapse due to the captive population exceeding the carrying capacity of their facility. Consequently, the Toledo zoo contacted other zoos in the United States about their willingness to house a captive Kihansi Spray Toad population. Between May and September 2010, the Toledo zoo transferred 500 animals to the Omaha zoo and 350 animals to both the Chattanooga and Detroit zoos. Yet this transfer of 1200 animals to other zoos in the United States was still insufficient to reduce the risk of the captive population exceeding the Toledo zoo's carrying capacity. Thus between July 2010 and September 2012, the Toledo zoo held back from rearing 12,352 births (Odum 2018).

With the establishment of two captive breeding facilities in Tanzania between 2010 and 2012 and the re-introduction of Kihansi Spray Toad back to the Kihansi Gorge in late 2012, the risk of a sudden population collapse of the captive population in the Toledo zoo was resolved through the regular transfer of animals back to Tanzania.

10.5 Water Quality Issues

Although by the late 2000s the Toledo and Bronx zoos had largely resolved the major animal husbandry challenges for the Kihansi Spray Toad, problems related to water quality did emerge over short periods of time in both zoos which resulted in short-term declines in the captive populations (Fig. 10.3) and number of births (Fig. 10.4).

Prior to 2014, the water source for the captive Kihansi Spray Toad population in the Toledo zoo was the city of Toledo's water supply which in turn was Lake Erie. In August 2014, there was a large algae bloom in Lake Erie that was attributed to agricultural runoff. In response to this algae bloom, the city of Toledo began to filter and purify its water using carbon, chlorine, and flocculants. This treatment resulted in the release of phosphates into the city water supply. In amphibians, phosphates can inhibit calcium metabolism.

In November 2014, the Toledo zoo installed a reverse osmosis water purification system to treat all of the water in their amphibian exhibits as well as within the biosecure captive breeding facility for the Kihansi Spray Toad (Andrew Odum *personal communication*). Following the installation of the reverse osmosis water purification system, the captive Kihansi Spray Toad population in the Toledo zoo rebounded within 6 months.

The Bronx zoo also experienced similar problems related to water quality. Between December 2007 and October 2008, the Kihansi Spray Toad population in their captive breeding facility declined from 436 to 282 animals. This decline in numbers of spray toads was traced also to phosphates in their water source. In late 2008, the Bronx zoo installed a new water filtration system for their captive Kihansi Spray Toad population which also addressed the short-term population decline of the Kihansi Spray Toad in their facility (Borek 2008).

References

Berger L, Speare R, Daszak P et al (1998) Chytridiomycosis causes amphibian mortality associated with population declines in the rain forests of Australia and Central America. PNAS 95(15):9031–9036

Borek A (2008) Kihansi Spray Toad report no. 15. Wildlife Conservation Society, Bronx, New York

Bronx Zoo (2011) Bronx Zoo Kihansi Spray Toad population. Annual report to National Environmental Management Council

Bronx Zoo (2014) Kihansi Spray Toad (Nectophrynoides asperginis). Number of groups and number of animals currently in the bio-secure facility at the Bronx zoo

Bronx Zoo (2017) Bronx Zoo Kihansi Spray Toad population. Annual report to National Environmental Management Council

Bronx Zoo (2019) Bronx Zoo Kihansi Spray Toad population. Annual report to National Environmental Management Council

Channing A, Finlow-Bates KS, Haarklau SE, Hawkes PG (2006) The biology and recent history of the critically endangered Kihansi Spray Toad Nectophrynoides asperginis in Tanzania. J East African Nat Hist 95:117–138. https://doi.org/10.2982/0012-8317(2006)95[117:tbarho]2.0.co;2

Chastain J (2019) Kihansi Spray Toad captive population. Excel database. Toledo Zoo, Toledo, Ohio

Herman T (2010) What can zoos contribute to amphibian conservation? PowerPoint Presentation

Lee S (2002) Kihansi Spray Toad report no. 4. Wildlife Conservation Society, Bronx, New York

Lee S (2003) Kihansi Spray Toad report no. 8. Wildlife Conservation Society, Bronx, New York

Lee S, Zippel K, Ramos L, Searle J (2006) Captive-breeding programme for the Kihansi spray toad Nectophrynoides asperginis at the Wildlife Conservation Society, Bronx, New York. Int Zoo Year 40:241–253. https://doi.org/10.1111/j.1748-1090.2006.00241.x

Odum A (2018) Kihansi Spray Toad captive population. Excel database. Toledo Zoo, Toledo

Planning for the Reintroduction of the Kihansi Spray Toad

11

Abstract

In late 2007, the Lower Kihansi Environmental Management Project and the IUCN Species Survival Breeding Specialist Group organized a population and habitat viability assessment workshop for the Kihansi Spray Toad. The objective of this workshop was to develop a management plan for the captive Kihansi Spray Toad population in US zoos and provide input for a species reintroduction plan for the spray toad. Because of concerns about potentially transmitting exotic pathogens back to Tanzania, it was agreed that biosecure captive breeding facilities would be established at the Bronx and Toledo zoos. It was also agreed at the population and habitat viability assessment workshop that two biosecure facilities would be constructed in Tanzania for further screening of pathogens and that the Bronx, Toledo, and San Diego zoos would develop a training program for Tanzanian scientists and animal keepers in establishing and managing amphibian biosecure captive breeding facilities. In 2010, LKEMP and the IUCN/SSC Amphibian Specialist Group organized a second international workshop that focused on the development of reintroduction guidelines for the Kihansi Spray Toad and the identification of key activities that would be required prior to transferring animals back to Tanzania including the development of a training program for Tanzanian scientists in amphibian pathology, microbiology, and molecular diagnostics.

11.1 Population and Habitat Viability Assessment Workshop

In October 2007, LKEMP and the IUCN Species Survival Commission's Conservation Breeding Specialist Group (CBSG) organized a population and habitat viability assessment workshop for the Kihansi Spray Toad. The objectives of this

W. D. Newmark, *The Kihansi Spray Toad* (Nectophrynoides asperginis): *Crisis Biology in the 21st Century*, Ecological Studies 254, https://doi.org/10.1007/978-3-031-89599-9_11

workshop were to develop a management plan for the captive population in the US zoos and provide input for a species reintroduction plan for the Kihansi Spray Toad.

Sixty individuals attended the workshop including representatives from academic and governmental institutions in Tanzania; IUCN; CBSG; Toledo, Bronx, and San Diego zoos; State University of New York, Syracuse; North-west University and University of the Western Cape in South Africa; environmental consulting firms (Hydrosphere and Afribugs); and the World Bank. The workshop was divided into five working groups that focused on the (1) cause of the sudden decline of the Kihansi Spray Toad in wild; (2) habitat change in the gorge; (3) management of the Kihansi Spray Toad in captivity and disease; (4) population modeling of the extinction risk of the Kihansi Spray Toad in captivity and in the wild; and (5) project management, organization, and resource conflicts (CBSG 2008).

An important outcome of this 4-day workshop was an agreement by participants that a program should be initiated to reintroduce the Kihansi Spray Toad back to the Kihansi gorge. This would require (1) establishing biosecure captive breeding facilities for the Kihansi Spray Toad at the Bronx and Toledo zoos; (2) expanding the size of the captive population in US zoos; (3) managing the Kihansi Spray Toad population at the Bronx and Toledo zoos to minimize loss of genetic diversity; (4) increase the screening of the Bronx and Toledo zoo populations for pathogens; (5) establish two captive breeding facilities in Tanzania to hold a captive population in-country for purposes of conducting additional screening for pathogens and challenge experiments with native species prior to re-introducing the Kihansi Spray Toad back to the gorge; and (6) training Tanzanians to manage a captive Kihansi Spray Toad population and screen, diagnose, and treat pathogens in a captive population (CBSG 2008).

Although the US zoos were initially concerned about transferring any Kihansi Spray Toads back to Tanzania because of the risk of accidently introducing exotic pathogens, the issue was resolved by an agreement that the two US zoos holding the captive Kihansi Spray Toad populations would construct biosecure facilities at their institutions to rear the Kihansi Spray Toad in isolation from other amphibian species housed in their collections. These biosecure facilities were patterned after biosecure captive breeding facilities that had been designed and tested in Australia.

In early 2008, the Bronx zoo established a biosecure facility for the Kihansi Spray Toad by retrofitting a shipping container. This was followed several years later by dedicating an isolated room in the Bronx zoo as a biosecure facility (Fig. 11.1). In 2008, the Toledo zoo also dedicated an isolated room in their Amphibian House as a biosecure facility (Fig. 11.2). Within both of these biosecure facilities, newborn toads were separated from the adult population, and animals were regularly screened for multiple diseases with the objective of establishing a disease-free population that could be transferred back to Tanzania.

The creation of these biosecure captive breeding facilities for the Kihansi Spray Toad at the Bronx and Toledo zoos was the first such facilities for any amphibian species held in US zoos and has now served as a model for managing amphibian species for reintroduction that have been held outside of their natural range in US zoos (Pessier and Mendelson 2010).

Fig. 11.1 The biosecure facility housing the Kihansi Spray Toad at the Bronx zoo. From left to right are Amina Kibola, project coordinator for the Kihansi Catchment Conservation and Management Project; Avi Shuter, Kihansi Spray Toad attendant; William Newmark and Jane Kibbassa, World Bank task team project leader. (Photo Credit: Amina Kibola)

Fig. 11.2 The biosecure facility constructed for the Kihansi Spray Toad at the Toledo zoo. From left to right are Andrew Odum, curator of herpetology, and Timothy Herman, Kihansi Spray Toad attendant standing in front of the public viewing window of the biosecure facility

11.2 Establishment of Captive Breeding Facilities in Tanzania

In early 2008, the Bronx and Toledo zoos initiated a training program for Tanzanian scientists and animal keepers in the establishment and management of amphibian biosecure captive breeding facilities. At the PHVA workshop, it was proposed that two captive breeding facilities be established in Tanzania: one at the University of Dar es Salaam and the second at the base of the Kihansi Gorge. Two Tanzanian scientists, Charles Msuya and Wilirk Ngalason from the University of Dar es Salaam, attended in late 2008 the first training course organized by the Bronx and Toledo zoos. This course focused on the design and management of a biosecure facility, including the required plumbing, electrical backup and sanitary systems, animal health care and diagnostics, culturing and management of insect food colonies, and data collection and storage.

Over the following 2 years, seven Kihansi Spray Toad attendants from Tanzania attended courses at the Bronx and Toledo zoos in the management of captive amphibian populations as well as data collection and management. Advanced training in amphibian disease diagnosis was also provided to faculty members from Sokoine University of Agriculture in Tanzania. In May 2010, Joshua Malago (Fig. 9.2) attended training courses at the San Diego and Bronx zoos in amphibian histopathology, and Gerald Misinzo (Fig. 11.3) received training in Che Weldon's lab at North-west University in South Africa in diagnosis of chytridiomycosis using in vitro cultivation and real-time PCR.

Fig. 11.3 Gerald Misinzo, a microbiologist at Sokoine University of Agriculture, received training in 2010 at North-west University in South Africa in diagnosis of chytridiomycosis using in vitro cultivation and real-time PCR analysis and conducted pre- and post-reintroduction testing for chytrid fungus in the Kihansi Spray Toad and in the Kihansi Gorge

In addition, an exchange program was established between US and Tanzanian captive breeding facilities for the Kihansi Spray Toad attendants. This exchange program was designed not only to provide training for Tanzanian attendants but to also allow attendants from US zoos to visit and assist their Tanzanian counterparts in establishing the captive breeding facility at the University of Dar es Salaam (UDSM).

In May 2009, construction of the biosecure captive breeding facility at the University of Dar es Salaam (Fig. 11.4) was completed. Shortly thereafter spray toad attendants, Alyssa Borek from the Bronx zoo and Timothy Herman from the Toledo Zoo, arrived in Tanzania and assisted UDSM scientists and Kihansi Spray Toad attendants in outfitting the captive breeding facility and establishing a food colony for the spray toads which consisted of *Drosophila* (fruit flies), Collembola (springtails), and pin-head crickets.

In mid-2010, three surrogate anuran species (southern torrent frog *Arthroleptides yakusini*, Uluguru tree frog *Leptopelis uluguruensis*, Tornier's forest toad *Nectophrynoides tornieri*) were collected from the Kihansi Gorge for purposes of testing the captive breeding facility and the staff's readiness to house the Kihansi Spray Toad. Several challenges were identified. First, raising sufficient food supplies for three species was problematic. This was resolved through expanding the collection and production of crickets from the wild. Secondly the population of *Nectophrynoides tornieri*, one of the three surrogate species, collapsed possibly due to an endo-parasite (Weldon et al. 2010), which illustrated the challenges of simultaneously managing multiple species in captivity.

Fig. 11.4 University of Dar es Salaam captive breeding facility

Fig. 11.5 (**a**) The Kihansi captive breeding facility is located at the base of the Kihansi Gorge. (**b**) Cages within this facility for housing the Kihansi Spray Toad

In May 2011, construction of the Kihansi Captive Breeding facility located at the base of the Kihansi Gorge was completed (Fig. 11.5). This captive breeding center was constructed by converting and outfitting several rooms in the Kihansi Visitors Center. The three attendants responsible for managing this facility had been trained at the Toledo and Bronx zoos.

11.3 Reintroduction Guidelines Workshop

In response to the consensus of the 2007 Population and Habitat Viability Assessment workshop that a reintroduction plan should be developed for the Kihansi Spray Toad, LKEMP in collaboration with IUCN/SSC Amphibian Specialist Group organized in February 2010 a workshop to develop reintroduction guidelines for the Kihansi Spray Toad. An important objective of the workshop was to identify pre-reintroduction activities and a timeline for their implementation prior to transferring the Kihansi Spray Toad back to the Kihansi Gorge.

The workshop was attended by representatives from academic institutions and government agencies in Tanzania as well as the IUCN Species Survival Commission; the Bronx, Toledo, and San Diego zoos; Amphibian Ark; Museo Tridentino di Scienze Naturali; North-west University; State University of New York, Syracuse; USGS Patuxent Wildlife Research Center; and the World Bank.

The key activities that the workshop identified that would be required prior to transferring any animals back to Tanzania were (1) careful screening for pathogens in the captive Kihansi Spray Toad population in the Bronx and Toledo zoos; (2) training Tanzanian scientists in amphibian pathology and in microbiology and molecular diagnostics; and (3) fully equipping and operationalizing the two Tanzanian captive breeding facilities including increasing the production of food items for the captive Kihansi Spray Toad population (McAloose et al. 2011).

It was also highlighted that after the transfer of a population of the Kihansi Spray Toad from US zoos to the UDSM biosecure captive breeding facility, an additional series of disease transmission experiments would be required that would involve exposing amphibian species and their substrates from the Kihansi Gorge to the US captive-reared spray toads.

The workshop also developed a detailed timeline for all activities that would be required prior to the release of the Kihansi Spray Toad back to the gorge. This timeline defined not only the order that activities would need to be implemented but also which organizations and individuals would be responsible for implementation.

Because the Kihansi Spray Toad in the US zoos had been initially housed with other species in cosmopolitan collections, a battery of tests and necropsies would be required to ensure that animals transferred back to Tanzania would not accidently introduce a novel pathogen back to Tanzania. This would entail conducting both a retrospective assessment of necropsy and histopathology data from the captive US Kihansi Spray Toad population, as well as additional screening of the captive US population for pathogens that must be assessed prior to acquiring the necessary international export and import permits.

Dee McAloose, a pathologist at the Bronx zoo, conducted the retrospective assessment. This assessment summarized the results of 867 necropsy submissions of which 556 animals were suitable for histologic review for the following pathogens: chytrid, ranavirus, enteric parasites, respiratory parasites, and intravascular ciliates. These pathogens had been previously detected in either the US captive Kihansi Spray Toad population or in the US institutions housing the Kihansi Spray Toad.

Screening for chytrid, ranavirus, enteric parasites, and intravascular ciliates was also a requirement for the acquisition of the necessary export and import permits. This screening was conducted by a team of pathologists and veterinarians including Allan Pessier from the San Diego zoo, Chris Hanley from the Toledo zoo, and Dee McAloose from the Bronx zoo.

The Kihansi Spray Toad Reintroduction Guidelines also outlined a series of experiments that should be initiated after the completion of the disease transmission experiments to assess whether the Kihansi Spray Toad could survive in the spray wetlands. The first proposed experiment was a soft release experiment in which the Kihansi Spray Toad would be housed at varying densities in cages placed in one or more of the spray wetlands to assess whether the reintroduced Kihansi Spray Toad could survive in the spray wetlands. If the Kihansi Spray Toad did not survive in the spray wetlands, then tests for potential stressors including chytrid fungus should be conducted through histopathological and PCR analysis. Additionally, the Reintroduction Guidelines proposed a series of hard releases of one or more populations of the Kihansi Spray Toad into one or more of the spray wetlands. Lastly, the Kihansi Spray Toad Reintroduction Guidelines highlighted the importance of concurrently monitoring key ecological components in the Kihansi Gorge including the spray wetland microclimate, water quality, vegetation, invertebrates, other amphibians, and predators following the established monitoring protocols.

References

CBSG (2008) Population and habitat viability assessment for the Kihansi Spray Toad. Apple Valley, MN

McAloose D, Magurisha H, Hanley C et al (eds) (2011) Kihansi Spray Toad re-introduction guidelines, Arlington

Pessier AP, Mendelson JR (eds) (2010) A manual for control of infectious diseases in Amphibian survival assurance colonies and reintroduction programs. IUCN/SSC Conservation Breeding Specialist Group. Apple Valley

Weldon C, Charles M, Gerald M, Malago J (2010) Ex-situ disease transmission experiment. Research proposal. Kihansi Spray Toad reintroduction program, Lower Kihansi Environmental Management Project

Establishment of Captive Breeding Populations in Tanzania

12

Abstract

The Toledo and Bronx zoos transferred 100 Kihansi Spray Toads to the University of Dar es Salaam captive breeding facility in August 2010. In early 2011, two disease transmission experiments were performed. In the first experiment, three development stages—larvae, metamorphs, and adults—of *Phrynobatrachus acridoides*, a native amphibian species that occurs in the Kihansi Gorge, were housed with the Kihansi Spray Toad. At the end of the experiment, all animals were swabbed for chytrid fungus and subjected to a post-mortem analysis which was negative. In the second set of experiments, Kihansi Spray Toads were housed at three densities in cages for 2 weeks that contained spray wetland vegetation and soil collected from potential reintroduction sites in the Kihansi Gorge. At the end of the experiment, animals were swabbed and tested from chytrid fungus which was negative. In June 2011, LKEMP formally ended. A Global Environment Facility supported project, the Kihansi Catchment Conservation and Management Project was designed during mid-2011 as a follow-on project.

Although by mid-2010, the captive breeding center at the University of Dar es Salaam was prepared to accept the Kihansi Spray Toad, transferring CITES 1 species between countries is not easy administratively, and the processing time for permits can be lengthy. Additionally, the Bronx, Toledo, and San Diego zoos wanted to exercise extreme caution in screening for pathogens in the captive population in the US zoos. The final tests that were performed by Allan Pessier from the San Diego zoo were screening for *Ranavirus* and *Batrachochytrium dendrobatidis*, which were negative.

W. D. Newmark, *The Kihansi Spray Toad* (Nectophrynoides asperginis): *Crisis Biology in the 21st Century*, Ecological Studies 254, https://doi.org/10.1007/978-3-031-89599-9_12

In August 2010, 100 Kihansi Spray Toads from the Toledo and Bronx zoos were transferred to the University of Dar es Salaam captive breeding facility. Detailed protocols had been prepared in advance by Fadhila Khatibu, the LKEMP project coordinator, as well as the University of Dar es Salaam and Sokoine University of Agriculture on receiving the Kihansi Spray Toad at the Dar es Salaam airport and transporting animals to the UDSM captive breeding facility. Alyssa Borek, the Kihansi Spray Toad attendant at the Bronx Zoo, accompanied the spray toads on their return flight to Tanzania and brought with her additional supplies and equipment for the UDSM captive breeding facility.

The Kihansi Spray Toad's arrival at the international airport in Dar es Salaam on 11 August was met by LKEMP's project coordinator, Fadhila Khatibu, as well as representatives from UDSM, SUA, and Ministry of Natural Resources and Tourism and members of the Tanzanian press. Over the following month, a series of popular articles appeared in Tanzanian papers and international periodicals highlighting the return of the Kihansi Spray Toad to Tanzania. This was a proud moment for not only Tanzania but also for the Bronx, Toledo, and San Diego zoos that had worked so hard over the previous 12 years to manage the captive Kihansi Spray Toad population, screen for potential pathogens, and train Tanzanian counterparts.

12.1 Disease Transmission Experiments

In the guidelines that were developed during the February 2010 Reintroduction Workshop, it was agreed that one or more disease transmission experiments would need to be conducted prior to the reintroduction of the Kihansi Spray Toad back to the wild. The objective of the disease transmission experiments was to determine whether it would be safe to expose native amphibian species occurring in the Kihansi Gorge to captive-reared Kihansi Spray Toads.

In early 2011, Ché Weldon, Charles Msuya, Gerald Misinzo, and Joshua Malago initiated a 2-month experiment in which the Kihansi Spray Toad was housed with a native amphibian species that occurs in the Kihansi Gorge. Three anuran species that occur in the Kihansi Gorge *Petropedetes yakusini*, *Nectophrynoides tornieri*, and *Leptopelis uluguruensis* had been collected and transferred to the UDSM captive breeding center in mid-2010 to test the captive breeding facility and husbandry protocols. However, due to issues related to abundance in captivity, ease of husbandry, and predation risk to the Kihansi Spray Toad, none of these species were selected for the disease transmission experiment. Rather, *Phrynobatrachus acridoides*, which also occurs in the Kihansi Gorge and in the Kilombero flood plain below the Kihansi gorge, was chosen (Weldon et al. 2011).

The first set of experiments conducted over 1 month involved housing in separate trials three development stages (larvae, metamorphs, and adults) of *Phrynobatrachus acridoides* with the Kihansi Spray Toad with appropriate controls. Prior to the experiment, *P. acridoides* was tested for chytrid fungus and treated with itraconazole baths to eliminate the possibility of transmission of chytrid fungi from *P. acridoides* to the Kihansi Spray Toad during the experiment. At the end of the experiment, all of the animals were euthanized, swabbed for chytrid fungi, and subjected to a post-mortem examination. All of the results of the post-mortem analyses were negative (Malago 2012).

The second set of experiments was designed to assess whether spray wetland vegetation and soil at potential reintroduction sites could potentially transmit chytrid fungus to the Kihansi Spray Toad. In this experiment, which was conducted by Charles Msuya, Gerald Misinzo, and Severinus Mutagwaba, Kihansi Spray Toads were housed at three densities (5, 10, and 20 animals) in cages with appropriate controls for 2 weeks that contained spray wetland vegetation and soil collected from the Kihansi Gorge. One day before the experiment began as well as at the end of the experiment and every 5 days during the experiment, all animals and cages were swabbed and tested for chytrid fungus using real-time PCR. Individual toad weight and snout-vent length were also recorded at the beginning and end of the experiment. All swabs for chytrid fungus from animals in cages containing sterilized and non-sterilized vegetation and soil were negative, and snout-vent length of animals did not differ between treatments and controls (Msuya et al. 2012). The results of these two disease transmission experiments thus provided a green light to proceed with the reintroduction of the Kihansi Spray Toad back to the Kihansi Gorge.

12.2 Securing Additional Funding for the Reintroduction

At the end of June 2011, LKEMP formally ended. Reintroducing the Kihansi Spray Toad back to the gorge would require securing additional support for the continuation of the species recovery program for the Kihansi Spray Toad. Over the next year and a half, Jane Kibbassa, Rafik Hirji, and I developed a follow-on 5-year project proposal that was submitted to the Global Environment Facility (GEF). The attractive feature of a GEF project particularly in the eyes of the Tanzanian government was that the project would be a grant rather than a loan.

The Tanzanian government rated the Kihansi proposal as its highest priority among competing GEF proposals that had been submitted from Tanzania. A condition, however, of GEF support was that conservation activities would need to be expanded to cover the other threatened species occurring in the Kihansi Gorge.

The objective of the GEF-supported Kihansi Catchment Conservation and Management Project (KCCMP) was to enhance biodiversity conservation in the Kihansi catchment and comprised of two major components:

1. *Institutional capacity building for the management of the Kihansi catchment*

   This component focused largely on the development of operational guidelines for conducting Environmental Flow Assessments so as to permit regulatory governmental organizations to address downstream impacts of water resource infrastructure projects in future Environmental Impact Assessments and the development of a long-term financing plan for the management of the Kihansi catchment.

2. *Conserve endangered species in the Kihansi catchment*
   This component focused largely on the reintroduction of the Kihansi Spray Toad back to the Kihansi Gorge and assessing the status in the Kihansi catchment of the newly described coffee (*Coffea kihansiensis*) (Davis and Mvungi 2004) and butterfly species (*Charaxes mtuiae*) (Collins et al. 2017). Additional activities under this component included training upstream communities in the Kihansi catchment in integrated pest management and mapping permanent and perennial water sources in the Kihansi catchment.

The Kihansi Catchment Conservation and Management Project (KCCMP) was initiated in December 2013 and continued through the end of December 2019. However, given the necessity of continuing to support the captive Kihansi Spray Toad populations in the United States and Tanzania and maintaining the spray irrigation systems, as well as continuing the ecological monitoring of key indicators in the Kihansi Gorge, it was important that bridging support be found to cover these activities between end of June 2011 when LKEMP ended and December 2013 when KCCMP would begin.

In mid-2011, the World Bank was providing support to Tanzania under the Tanzania Energy Development and Access Expansion Project (TEDEP) that was focused on abating greenhouse gas emissions through the use of renewable energy in rural areas. One objective of this project was to strengthen TANESCO's capacity in environmental management. Because of TANESCO's legal obligation to mitigate the environmental impacts associated with the Lower Kihansi hydropower facility, it was agreed that 18 months of bridging support would be provided under TEDEP to initiate the reintroduction of the Kihansi Spray Toad back to the Kihansi Gorge, as well as maintain the captive Kihansi Spray Toad populations and the gorge infrastructure and continue the ecological monitoring of key indicators in the Kihansi Gorge.

References

Collins S, Congdon C, Bampton I (2017) Review of the Charaxes gallagheri complex resulting in two new species and undescribed morphs of Charaxes gallagheri van Son, 1962 (Lepidoptera, Nymphalidae, Charaxinae). Entomol Africana 22:19–30

Davis AP, Mvungi EF (2004) Two new and endangered species of Coffea (Rubiaceae) from the Eastern Arc Mountains (Tanzania) and notes on associated conservation issues. Bot J Linn Soc 146:237–245. https://doi.org/10.1111/j.1095-8339.2004.00328.x

Malago J (2012) Report on disease transmission experiments. Sokoine University of Agriculture. Morogoro, Tanzania. Lower Kihansi Environmental Management Project

Msuya CA, Misinzo G, Malago J (2012) Ex-situ exposure of Kihansi Spray Toad to Kihansi Gorge Spray wetlands substrate. Lower Kihansi Environmental Management Project

Weldon C, Msuya C, Misinzo G, Malago J (2011) Ex-situ disease transmission experiment. Lower Kihansi Environmental Management Project

13 Pre-reintroduction Assessment of the Spray Wetlands

Abstract

In mid-2006, LKEMP commissioned NORPLAN to install a backup spray irrigation system in the Upper, Lower, and Mid-Gorge Spray Wetlands in case the primary spray irrigation systems were to fail. Due to the improvement in the volume and dispersion of spray that was provided by the newly installed backup spray irrigation system, the backup spray irrigation system became the primary spray irrigation systems in these three spray wetlands. However, by mid-2009 due to declining water pressure again in the primary and backup spray irrigation systems, the two spray irrigation systems were jointly operated. As a result of the installation of the primary and backup sprinkler systems, plant community structure in the Upper Spray Wetland recovered sufficiently following commissioning of the hydroelectric facility that a recommendation was made to proceed with the reintroduction of the Kihansi Spray Toad to the Kihansi Gorge. The monitoring of air temperature and relative humidity within the Upper, Lower, and Mid-Gorge Spray Wetlands during 2010 shortly before the reintroduction of the Kihansi Spray Toad found that mean daily temperature and relative humidity fell within the range that had been recorded within these three wetlands in 2001. Lastly, the monitoring of the abundance of *Afrosteles distans*, one of the two most important food items in the diet of the Kihansi Spray Toad, revealed that *A. distans'* abundance had remained stable between 2001 and 2018, thus also providing a green light to proceed with the reintroduction of the Kihansi Spray Toad.

W. D. Newmark, *The Kihansi Spray Toad* (Nectophrynoides asperginis): *Crisis Biology in the 21st Century*, Ecological Studies 254, https://doi.org/10.1007/978-3-031-89599-9_13

13.1 Spray Wetland Vegetation

Since the commissioning of the Kihansi hydroelectric facility in December 1999 which had resulted in a 83% reduction in the area of the spray wetlands receiving natural spray, there was concern about whether the spray wetland vegetation could recover sufficiently to provide adequate habitat for the Kihansi Spray Toad. This concern was highlighted during the PHVA workshop in October 2007. The monitoring of spray wetland vegetation was identified as a top-tier indicator in the ecological monitoring protocols that were developed for the Kihansi Gorge ecosystem (Gibbs 2006).

Between June 2000 and March 2001, gravity-fed spray irrigation systems were constructed over portions of the Upper, Lower, and Mid-Gorge Spray Wetlands. These spray irrigation systems provided an input of >70 mm per day. However, the spray irrigation systems in the Upper, Lower, and Mid-Gorge Spray Wetlands required regular maintenance, and therefore having a backup spray irrigation system was important in case the primary spray irrigation systems were damaged by flooding, landslides, or other natural disasters.

Between 2001 and 2012, there were four incidences of the gravity-fed sprinkler systems being damaged by high water, flooding, or landslides. Fortunately, a highly experienced maintenance team responsible for the weekly maintenance of the spray irrigation systems was able to repair the spray irrigation systems in less than 2 days with the exception of damage that occurred in 2008 as a result of unusually high flows in the Kihansi River. During March and April 2008, coinciding with the long rains in south-central Tanzania, the spray irrigation systems in the Upper and Lower Spray Wetlands were extensively damaged due to flows in the Kihansi River that exceeded 21 m^3/s. Following the decline in flow of the Kihansi River at the end of April 2008, the maintenance team was able to repair the two spray irrigation systems over 10 days. However, the recurring need to maintain and repair the sprinkler systems in the Kihansi Gorge highlighted the importance of having a backup sprinkler system.

13.1.1 Installation of the Backup Sprinkler System

In mid-2006, LKEMP commissioned NORPLAN to install a backup spray irrigation system in the Upper, Lower, and Mid-Gorge Spray Wetlands. This involved installing a second set of piping and nozzles that ran parallel and adjacent to each of the original lines in the three wetlands.

The combination of higher pressure and new nozzles greatly increased the volume and dispersion of spray over the wetlands. As a result of this improvement, a decision was made to operate the newly installed backup spray irrigation system as the primary spray irrigation system and the older original primary spray irrigation system as the backup. However, by mid-2009 the water pressure in the newly installed backup spray irrigation system also began to decline, and in December

2009 a decision was made to jointly operate the primary and backup sprinkler systems in the Upper, Lower, and Mid-Gorge Spray Wetlands.

13.1.2 Response of the Spray Wetland Vegetation

Following the installation in 2006 of the backup spray irrigation systems in the Kihansi spray wetlands, there was a noticeable visual improvement in the physical structure of the spray wetland vegetation in the Upper and Lower Spray Wetlands. The percent cover of *Brillantaisia madagascariensis* and *Aframomum mala*, perennial herbs that grow to 1.5–2 m in height, declined in the Upper and Lower Spray Wetlands following the deployment of the new backup sprinkler system. Prior to the commissioning of the Kihansi dam, *Brillantaisia madagascariensis* and *Aframomum mala*,had been restricted largely to the edge of the spray wetlands (Quinn et al. 2005; CELP 2006). However following the commissioning of the hydroelectric facility, the percent cover of these two species noticeably increased within the interior of the Upper and Lower Spray Wetlands (Fig. 13.1).

The visual improvement in the physical structure of the Upper and Lower Spray Wetlands between 2006 and 2009 was confirmed by the annual monitoring of plant species in the Upper Spray Wetland. In April 2001, Jon Lovett and his graduate

Fig. 13.1 Following the commissioning of the Lower Kihansi hydroelectric facility, the percent cover of *Brillantaisia madagascariensis* and *Aframomum mala* greatly increased in the interior of the Upper Spray Wetland, as shown here in June 2004. Prior to the commissioning of the Lower Kihansi hydroelectric facility, these two plant species were largely restricted to the edge of the spray wetland

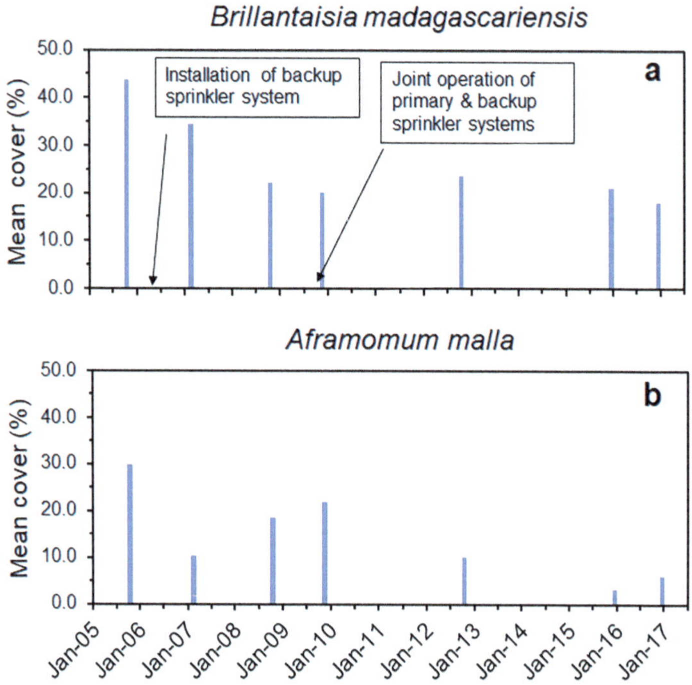

Fig. 13.2 Change in mean percent cover of (**a**) *Brillantaisia madagascariensis* and (**b**) *Aframomum mala* between October 2005 and December 2015 in the Upper Spray Wetland following the installation of the backup spray irrigation system in June 2006 and the joint operation of the primary and backup sprinkler systems in the Upper Spray Wetland in December 2009. (Source: Ndangalasi 2009, 2010, 2017a, b)

students from the University of York and Henry Ndangalasi from the University of Dar es Salaam established eight permanent plots in the Upper Spray Wetland. Six of the eight plots were placed in locations in the Upper Spray Wetland that received spray from the spray irrigation system, and two plots were placed in locations that did not received spray. All plots were monitored annually between 2001 and 2008 with the exception of 2006 (Fig. 13.2).

In 2014, Vigdis Vandvik from the University of Bergen and colleagues published a summary of the changes in the spray wetland vegetation between plots within the Upper Spray Wetland that received artificial spray in comparison to those that did not receive artificial spray (Vandvik et al. 2014). They reported that by 2008 portions of the Upper Spray Wetland that had received artificial spray had stabilized and that the wetland plant species that Alan Channing had identified as associated with the Kihansi Spray Toad (*Selaginella kraussiana*, *Pilea rivularis*, *Panicum hymeniochilum*, *P. parvifolium*, *Leersia hexandra*, *Impatiens* spp.) were increasing. They also noted that the improvement in the spray wetland vegetation had begun within 2 years after the installation of the spray irrigation system, and the stabilization of the spray wetland vegetation commenced over the following 6 years. Vigdis

Vandvik et al. concluded that the spray wetland vegetation in the Upper Spray Wetland had sufficiently recovered by 2008 to proceed with the reintroduction of the Kihansi Spray Toad.

In contrast to the noticeable improvement in the Upper and Lower Spray Wetland vegetation between 2001 and 2009, the spray wetland vegetation in the Mid-Gorge Spray Wetland had improved comparatively less over this same time period. In the Mid-Gorge Spray Wetland, *Brillantaisia madagascariensis* had continued to dominate the overstory in this spray wetland following the installation of the backup spray irrigation system (Fig. 13.3).

In 2005, Devolent Mtui (Fig. 13.4), the LKEMP research and monitoring program officer, initiated a series of weeding experiments following the recommendations of the External Review Panel, to assess whether manually removing overstory species could improve the abundance plant species that were declining in the Mid-Gorge Spray Wetland. She found that removal in experimental plots of all plant species with the exception of *Selaginella kraussiana* had a positive effect. The percent cover of *Selaginella kraussiana* increased after 2 months following this initial weeding.

She then examined between 2006 and 2008 the influence of manual weeding and soil composition on regeneration of plant species associated with the Kihansi Spray Toad. In these experiments she found that regeneration of *Panicum* spp. and *Selaginella kraussiana* occurred more rapidly in silty soils than in sandy soils. This result suggested that soil erosion that occurred in the Mid-Gorge Spray Wetland during the pilot testing of the large rotating spray gun in 1999 (Sect. 4.2) was quite

Fig. 13.3 The Mid-Gorge Spray Wetland vegetation in November 2000. *Brillantaisia madagascariensis* has dominated the overstory of this spray wetland since 2000. (Photo credit: TANESCO)

Fig. 13.4 Devolent Mtui served as the research and monitoring program officer in the Kihansi Gorge between 2004 and 2008 under LKEMP and since 2020 has lead the reintroduction of the Kihansi Spray Toad back to the Kihansi Gorge under TAWIRI

likely contributing to the slower recovery of the vegetation composition and structure in this wetland. Devolent Mtui concluded that improving the vegetation composition and structure in the Mid-Gorge Spray Wetland will likely require amendments of silty soils to this wetland (Mtui 2009).

13.1.3 Expansion of the Spray Irrigation Systems in the Spray Wetlands

In early 2011, LKEMP commissioned NORPLAN to expand the spray irrigation systems in the Upper and Lower Spray Wetlands and to reduce the droplet size generated by the fountain jet sprinkler systems situated along the Kihansi River adjacent to the Upper Spray Wetland (Fig. 6.2). The purpose of expanding the spray irrigation systems in the Upper and Lower Spray Wetlands was to further prevent the encroachment of overstory edge species and potential predators of the Kihansi Spray Toad—snakes and ants—from entering the Upper and Lower Spray Wetlands. Sixty additional nozzles were installed along 2 new spray irrigation lines on the eastern and southern edges in the Upper Spray Wetland, and 30 nozzles were installed along 1 new spray irrigation line along the eastern edge of the Lower Spray Wetland. Additionally, a steel plate was welded on the front of the two fountain jets which was successful in reducing the size of the droplets generated by the fountain jet sprinkler systems (NORPLAN 2011).

13.2 Microclimate and Water Quality

The Kihansi Spray Toad reintroduction guidelines also highlighted the importance of continuing to monitor the microclimate and water quality in the spray wetlands both prior and following the reintroduction of the Kihansi Spray Toad back to the Kihansi Gorge (McAloose et al. 2011). In 2001, following the installation of the spray irrigation systems in the Upper, Lower, and Mid-Gorge Spray Wetlands, mean daily temperature and relative humidity within the spray wetlands ranged between 15.7–20.7 C^0 and 80.0–98.8%, respectively (NORPLAN 2002).

Measurements conducted in 2010 by Severinus Mutagwaba, the LKEMP research and monitoring program officer, immediately prior to the reintroduction of the Kihansi Spray Toad back to the Kihansi Gorge found that mean daily temperature and relative humidity in the Upper, Lower, and Mid-Gorge Spray Wetlands fell within the mean daily range for temperature and relative humidity[1] that were recorded in 2001 in these three wetlands (Mutagwaba 2010).

In 2010, Severinus Mutagwaba initiated the monitoring of water quality in the spray wetlands and in the Kihansi Gorge. Water temperature, pH, dissolved oxygen, and conductivity were recorded weekly (Fig. 13.5) at eight locations in the gorge (dam intake, dam bypass, Jabali intake, Upper Spray Wetland sprinklers, Lower Spray Wetland sprinklers, Handaki intake, Daraja kubwa, and the Tailrace). These data form a baseline for the future monitoring of water quality in the Kihansi Gorge and spray wetlands.

The Rufiji Basin Water Board also monitored between 2013 and 2019 the pH of water and fecal coliform levels along three tributary streams flowing into the Kihansi Gorge—Mhalala stream (upstream of the gorge), Jibali stream (upstream of the gorge), and Handaki stream (within the gorge). Over this 7-year period, the pH of water in these three tributary streams remained relatively constant, while fecal coliform levels resulting from human settlements and livestock in the micro-catchments of these tributary streams fluctuated seasonally with higher fecal coliform levels during the long rains.

13.3 Spray Wetland Invertebrates

The Kihansi Spray Toad reintroduction guidelines also recommended the continuing monitoring within the spray wetlands of the two most common food items in the diet of the Kihansi Spray Toad, *Afrosteles distans* (Homoptera: Cicadellidae) and *Ortheziola* sp. (Hemiptera: Ortheziidae). However, due to turnover of personnel and issues related to funding, only *Afrosteles distans* was regularly monitored between 2001 and 2018 in the Upper Spray Wetland. This wetland was selected as a monitoring site for *Afrosteles distans* because the Upper Spray Wetland is the largest spray

[1] The exception being mean minimum relative humidity in the Mid-Gorge Spray wetland in June 2010 which was 52.5%. However, this very likely was a result of the location of the data logger in relation to the spray irrigation lines

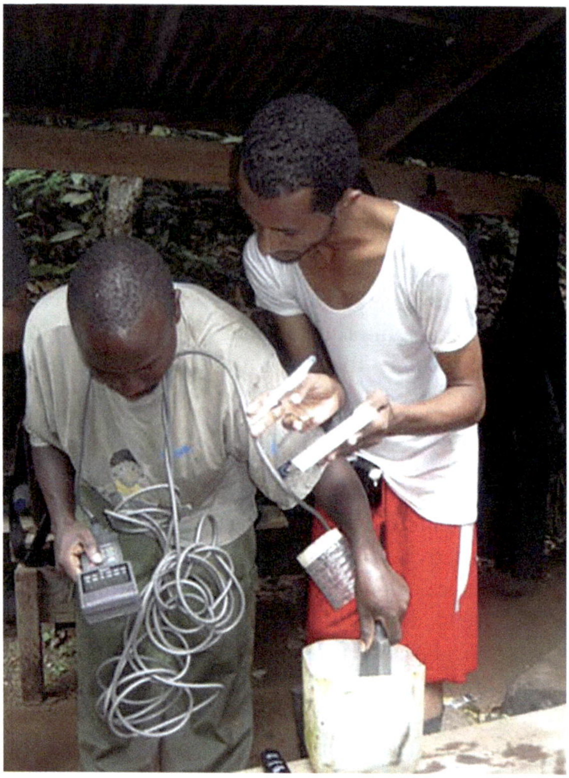

Fig. 13.5 Severinus Mutagwaba, the research and monitoring program officer between 2010 and 2014 under LKEMP and KCCMP, and Nassoro Mohamed, University of Dar es Salaam MSc student, monitoring water quality from water samples collected in the Upper Spray Wetland. (Photo credit: Kurt Buhlmann and Tracey Tuberville)

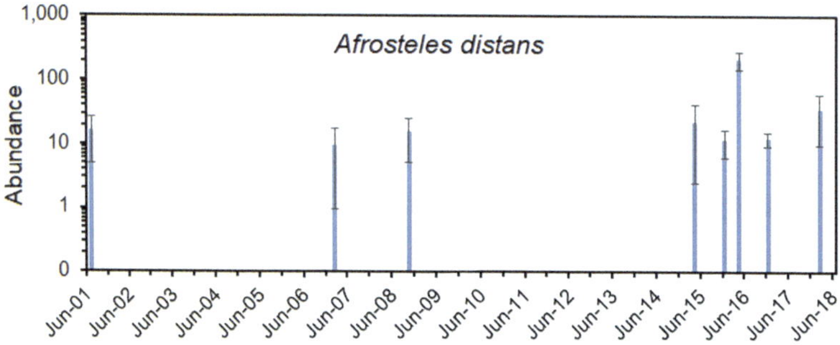

Fig. 13.6 Mean (±95% C.I.) abundance of *Afrosteles distans* across six plots in the Upper Spray Wetland between July 2001 and February 2018. The abundance of *A. distans* was assessed by placing a plastic dish containing 50 ml of 70% ethanol on the ground and raising and dropping a 50 cm square frame around the plastic dish three times to disturb the insects and then recording the number of *A. distans* that were captured in the plastic dish. Cicadellids jump when disturbed and a proportion land in the dish. Ten placements of the dish were used to collect one sample; and 10 drop-samples were collected per plot. (Source: NORPLAN 2001; Kaaya 2018)

wetland and had historically held the largest proportion of the total Kihansi Spray Toad population in the Kihansi Gorge and thus was identified as the primary reintroduction site for the Kihansi Spray Toad.

Between 2001 and 2018, the abundance of *Afrosteles distans* in the Upper Spray Wetland remained relatively stable (Fig. 13.6). The exception being April 2016 in which the abundance of *Afrosteles distans* increased 12-fold in comparison to mean abundance recorded over the other survey periods. The long-term monitoring of *Afrosteles distans* in the Upper Spray Wetland also revealed that changes in the operation of the primary and backup sprinkler systems in this spray wetland had minimal impact on its abundance.

References

CELP (2006) Lower Kihansi Hydropower Project. Lower Kihansi environmental monitoring program. Field report: Vegetation studies – October 2005. Report produced for the Lower Kihansi Environmental Monitoring Program, Dar es Salaam, Tanzania

Gibbs JP (2006) A detailed ecological monitoring protocol for the Lower Kihansi Gorge ecosystem. Final report to the Lower Kihansi Environmental Management Project

Kaaya L (2018) Macroinvertebrates survey and monitoring in the Kihansi Gorge during the dry season, February 2018. Kihansi Catchment Conservation and Management Project

McAloose D, Magurisha H, Hanley C et al (eds) (2011) Kihansi Spray Toad re-introduction guidelines. Arlington

Mtui D (2009) Ecological monitoring program: Progress report October 2008 to April 2009. Lower Kihansi Environmental Management Project

Mutagwaba S (2010) Ecological monitoring program: annual report December 2009–December 2010. Lower Kihansi Environmental Management Project

Ndangalasi HJ (2009) Draft field report. Kihansi Gorge: Upper spray wetland vegetation. Lower Kihansi Environmental Management Project

Ndangalasi HJ (2010) Draft field report. Kihansi Gorge: Upper spray wetland vegetation analysis. Lower Kihansi Environmental Management Project

Ndangalasi HJ (2017a) Monitoring of wetland vegetation in USW. Brief report: February 2017. Kihansi Catchment Conservation and Management Project

Ndangalasi HJ (2017b) Kihansi gorge (the upper and lower spray wetlands vegetation monitoring). October 2017. Kihansi Catchment Conservation and Management Project

NORPLAN (2001) Lower Kihansi Hydropower Project. Immediate rescue and emergency measures. Mid-term Report. Report produced for Tanzania Electric Supply Company Ltd (TANESCO), Dar es Salaam, Tanzania

NORPLAN (2002) Lower Kihansi hydropower project. Immediate rescue and emergency measures. Final specialist report: amphibian studies. Report produced for Tanzania Electric Supply Company Ltd. (TANESCO), Dar es Salaam, Tanzania

NORPLAN (2011) Assessment of technical options for primary and backup sprinkler system. Final Report. National Environment Management Council

Quinn CH, Ndangalasi HJ, Gerstle J, Lovett JC (2005) Effect of the Lower Kihansi Hydropower Project and post-project mitigation measures on wetland vegetation in Kihansi Gorge, Tanzania. Biodivers Conserv 14:297–308. https://doi.org/10.1007/s10531-004-5048-3

Vandvik V, Maren IE, Ndangalasi HJ et al (2014) Back to Africa: monitoring post-hydropower restoration to facilitate reintroduction of an extinct-in-the-wild amphibian. Ecosphere 5. https://doi.org/10.1890/ES14-00093.1

Reintroduction of the Kihansi Spray Toad

14

Abstract

With the assistance of IUCN and the IUCN/SSC Amphibian Specialist Group, LKEMP recruited two technical advisors to assist in initiating the reintroduction of the Kihansi Spray Toad back to the Kihansi Gorge. The advisers' first task was to design an experimental cage that would prevent newborn Kihansi Spray Toads from escaping but large enough to permit the entry and exit of small invertebrate food items. A 1 m × 1 m experimental cage with a 3 mm wire mesh was tested and selected. The IUCN consultants then conducted two replicated and one non-replicated experiments. In the first replicated experiment, adult survival was found to be higher in the Upper Spray Wetland than in Lower Spray Wetland. In the second replicated experiment, adult and sub-adult survival did not vary between the Upper and Lower Spray Wetlands although this result may have been affected by damage to the artificial spray irrigation system in the Upper Spray Wetland as a result of a landslide. In the third experiment, newborn survival was found to be 4.5 times higher than adult survival across the Upper and Lower Spray Wetlands.

14.1 Technical Assistance from IUCN

In June 2011, Fadhila Khatibu, LKEMP's project coordinator, contacted IUCN and requested their assistance in identifying a technical advisor(s) who could assist the project in initiating the reintroduction of the Kihansi Spray Toad back to the Kihansi Gorge. In response to this request, IUCN contacted Don Church, Director of IUCN/SSC Amphibian Specialist Group, who had helped organize the Kihansi Spray Toad reintroduction guideline workshop, to assist LKEMP in identifying potential candidates for this position.

W. D. Newmark, *The Kihansi Spray Toad* (Nectophrynoides asperginis): *Crisis Biology in the 21st Century*, Ecological Studies 254, https://doi.org/10.1007/978-3-031-89599-9_14

In October 2011, LKEMP, IUCN, and the World Bank selected Tracey Tuberville and Kurt Buhlmann, a couple from the University of Georgia Savannah River Ecology Laboratory who had extensive experience in the reintroduction of herpetofauna. Tracey Tuberville and Kurt Buhlmann were particularly well-qualified to lead this effort because of their experience in the design and implementation of soft-release reintroduction experiments. The importance of adopting an experimentally driven approach in the reintroduction of the Kihansi Spray Toad to the wild had been emphasized in project documents as well as in the reintroduction guidelines because such an approach can provide important insights into the mechanisms that may be hindering or contributing to failure or success of the reintroduction.

In April 2012, Tracey Tuberville and Kurt Buhlmann conducted a 10-day planning and reconnaissance visit to Tanzania to meet with Tanzanian counterparts from the University of Dar es Salaam, Sokoine University of Agriculture, and the Tanzania Wildlife Research Institute. During this initial visit, they conducted a site visit to the Kihansi Gorge and met Nassoro Mohamed a MSc student from the University of Dar es Salaam who they would be co-advising while in the field.

14.2 Pilot Cage Experiments

In June 2012, Tracey Tuberville and Kurt Buhlmann returned to Tanzania for 5 weeks. The objectives of this second visit were to design the experimental cages to house the Kihansi Spray Toad in the wetlands, test the use of visible elastomer implants as a marking technique for the Kihansi Spray Toad, initiate pilot experiments to assess whether a small reintroduced Kihansi Spray Toad population could survive in the cages in the spray wetlands, and design a series of laboratory experiments that would parallel the field experiments.

In late June 2012, Tracy and Kurt evaluated several proposed designs for experimental cages. The selected design for the experimental cage was 1 m x 1 m with a 3 mm wire mesh covering. This mesh was small enough to prevent any newborn Kihansi Spray Toads from escaping but large enough to permit the entry and exit of small invertebrate food items (Fig. 14.1).

Three cages were placed in the Upper Spray Wetland on leveled sites in the wetland, and soil and vegetation from the cleared sites were replaced in the base of the cages. Sixteen individually marked adults (8 males and 8 non-gravid females) were placed in Cage 1 to monitor survivorship; 24 non-marked sub-adults were placed in Cage 2 to monitor survivorship and growth; and 8 individually marked gravid-females were placed in Cage 3 to monitor reproductive success and newborn growth. The status of Kihansi Spray Toad in the cages was monitored every 2 weeks over 4 months. At the end of the experiment, all of the remaining spray toads were swabbed, and preserved samples were deposited with Gerald Misinzo at the Sokoine University of Agriculture and tested for chytrid fungus.

Results of these pilot experiments indicated that the cages were large enough to provide adequate habitat for the Kihansi Spray Toad and the mesh size of the screen covering the cages was small enough to prevent newborn Kihansi Spray Toad from

Fig. 14.1 Kurt Buhlmann, an IUCN consultant, placing vegetation into an experimental cage used in the caged experiments that were conducted between June and September 2012 in the Upper and Lower Spray Wetlands. (Photo Credit: Kurt Buhlmann and Tracy Tuberville)

escaping but large enough to allow small insects to enter the cage. Regularly cleaning the outside of the cages would, however, be required to prevent the blockage of the screen by algae and other debris.

These pilot studies suggested that survivorship of the Kihansi Spray Toad may possibly vary by age class. The percent subadults surviving after 4 months was approximately eight times higher than that for adults (50% vs 6.3%). The pilot experiments also indicated rapid development in newborn Kihansi Spray Toads. By month four, three of the newborn Kihansi Spray Toads in cage were patterned subadults. Finally, all of the skin swab samples tested negative for chytrid fungus (Buhlmann and Tuberville 2012; Tuberville and Buhlmann 2014).

14.3 Replicated Cage Experiments

Based on these preliminary findings, Tracey Tuberville and Kurt Buhlmann along with Nassoro Mohamed (Fig. 14.2) conducted two replicated and one non-replicated experiments to assess whether adult, subadult, and newborn survival vary between the Upper and Lower Spray Wetlands. The purpose of comparing survival between the Upper and Lower Spray Wetlands was to assess the comparative suitability of these two spray wetlands for subsequent reintroductions of larger numbers of Kihansi Spray Toads.

In the first experiment conducted between November 2012 and March 2013, three cages each containing 16 adult Kihansi Spray Toad that were individually marked were placed in the two wetlands. In the second experiment conducted between April and November 2013, three cages each containing 16 unmarked

Fig. 14.2 Nassoro Mohamed a University of Dar es Salaam MSc student and Kurt Buhlmann and Tracey Tuberville, IUCN consultants, also conducted laboratory experiments in the Kihansi captive breeding facility. (Photo credit: Kurt Buhlmann and Tracey Tuberville)

subadults were placed in Upper and Lower Spray Wetlands. In the third experiment also conducted between April and November 2013, ten newborn Kihansi Spray Toads were placed in a single cage in the Upper and Lower Spray Wetlands. In all of these experiments, toads were monitored every 2 weeks over a 4-month period and were swabbed and tested for chytrid fungus at the beginning and at the end of the experiments (Buhlmann and Tuberville 2012; Tuberville and Buhlmann 2014).

In the first experiment, daily adult survival after 4 months was seven times higher in the Upper than in the Lower Spray wetland (29.4% versus 4.2%) (Fig. 14.3). In the second experiment, daily sub-adult survival in the Upper Spray wetland was highly similar after 4 months to that observed in the Lower Spray Wetland. However, 2 days after the initiation of the second experiment, a landslide extensively damaged the spray irrigation system in the Upper Spray Wetland which altered the dispersion of artificial spray in the Upper Spray Wetland (Fig. 14.4), and thus quite likely reduced subadult survival in two of the three cages in the Upper Spray Wetland.

The third experiment revealed high and nearly identical survival of newborns in the Upper and Lower Spray Wetlands. After 4 months, survival of newborns in the Upper and Lower Spray Wetlands was 70% and 80%, respectively. Over the 4-month study period, mean snout to vent length of newborns nearly doubled, while body weight increased more than fivefold.

The third experiment also revealed that newborn survival after 125 days was 4.5 times higher than adult survival across the Upper and Lower Spray Wetlands (Fig. 14.5). All of the skin swabs from the Kihansi Spray Toad collected at the beginning and the end of the three experiments tested negative for chytrid fungus (Misinzo, *personal communication*).

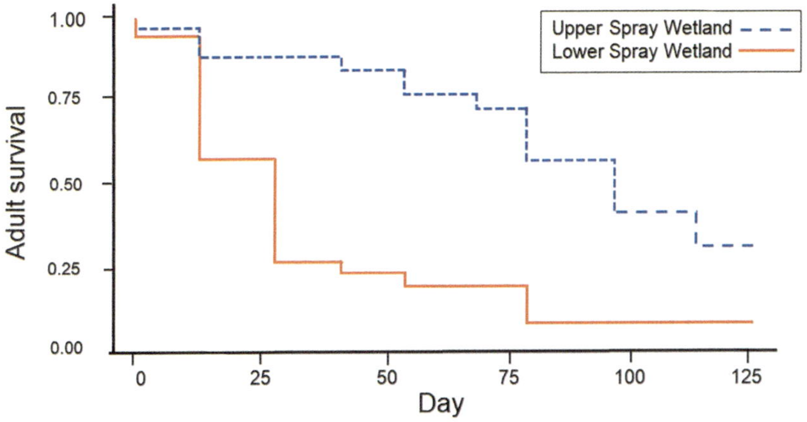

Fig. 14.3 Comparison of daily adult survival of the Kihansi Spray Toad between the Upper and Lower Spray Wetlands over 125 days based on a Kaplan-Meier survival analysis. (Source: Mohamed and Magige (2020))

Fig. 14.4 Damage to the Upper Spray Wetland and sprinkler system resulting from a landslide that occurred on 15 March 2013. (Photo credit: Kurt Buhlmann and Tracey Tuberville)

The results of these three experiments indicated that Kihansi Spray Toad survival in the Upper and Lower Spray wetland varied among age classes with highest survival among newborns, followed by subadults and then adults. Yet obviously what was unknown was how Kihansi Spray Toad survival would vary over a more

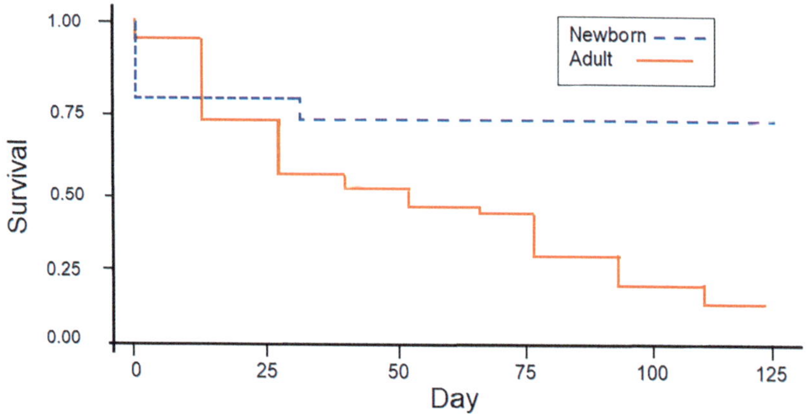

Fig. 14.5 Comparison of newborn and adult daily survival of the Kihansi Spray Toad across the Upper and Lower Spray Wetlands over 125 days based on a Kaplan-Meier survival analysis. (Source: Mohamed and Magige (2020))

extended timeframe. And thus, this set the stage for the first two hard releases of Kihansi Spray Toad in the Kihansi Gorge spray wetlands over the following 6 months.

References

Buhlmann KA, Tuberville TD (2012) Progress report: Kihansi Spray Toad reintroduction project. Lower Kihansi Environmental Management Project

Mohamed N, Magige FJ (2020) Survivorship of the reintroduced Kihansi Spray Toads Nectophrynoides asperginis in a soft release experiment at the Kihansi Gorge, Tanzania. J Educ Humanit Sci 9:111–122

Tuberville TD, Buhlmann KA (2014) Final report on Kihansi Spray Toad reintroduction efforts during 2012–2013, including progress report for December 2012–July 2013. Kihansi Catchment Conservation and Management Project

15 Hard Releases of the Kihansi Spray Toad Back to the Kihansi Gorge

Abstract

On 30 October 2012, 2000 Kihansi Spray Toads were reintroduced in a hard release to the Upper Spray Wetland. This reintroduction represented the culmination of 12 years of effort by multiple institutions and individuals to conserve the Kihansi Spray Toad in the wild. In early November 2012, 435 spray toads were also reintroduced to the Mhalala Spray Wetland. The abundance of the Kihansi Spray Toad in both spray wetlands declined rapidly over the first month followed by a slower decline over the following 2 to 3 months. A second hard release of 1500 animals was conducted in the Upper Spray Wetland in March 2013. Change over time in abundance of the Kihansi Spray Toad following the second release was highly similar to that observed during the first releases in the Upper and Mhalala Spray Wetlands—a rapid decline after the first month follower by a slower decline over the following 2 months. Nonetheless, an encouraging result emerged from the first two hard releases in the Upper Spray Wetland. A very small population survived up to 2 years in the Upper Spray Wetland following the first and second hard releases. The small surviving population contained gravid females and juveniles indicating post-release reproduction and survival of neonates. A third hard release was conducted in the Upper Spray Wetland in February 2015. Highly similar patterns were observed again in the change in abundance over time of the Kihansi Spray Toad with a very small reproducing population surviving up to 2 years in the Upper Spray Wetland following the third hard release.

W. D. Newmark, *The Kihansi Spray Toad* (Nectophrynoides asperginis): *Crisis Biology in the 21st Century*, Ecological Studies 254, https://doi.org/10.1007/978-3-031-89599-9_15

15.1 First Hard Release

On 30 October 2012, 2000 Kihansi Spray Toads were reintroduced back to the Upper Spray Wetland. This release represented the first reintroduction of an extinct-in-the-wild anuran species in the Afrotropics and only the second reintroduction of an extinct-in-the-wild anuran species worldwide. For the many individuals and institutions who had worked so tirelessly over the previous 12 years to conserve the Kihansi Spray Toad, this was a day for celebration.

The reintroduction of the Kihansi Spray Toad represented an important milestone in the species recovery program for the Kihansi Spray Toad. Considerable planning and coordination were required to transfer more than 2000 toads from captive breeding facilities in the United States to the Kihansi Gorge (Figs. 15.1 and 15.2). Prior to the release, the IUCN consultants met with the Tanzanian Kihansi Spray Toad reintroduction committee consisting of faculty and staff from the University of Dar es Salaam and Sokoine University of Agriculture as well as representatives from LKEMP and the Wildlife Division who reviewed the proposed reintroduction protocols. It was agreed that 2000 Kihansi Spray Toads would be released in the Upper Spray Wetland and 500 spray toads would be released in the Mhalala Spray Wetland. Because of concerns raised by the Tanzanian reintroduction committee about potential adverse effects of the elastomer tags on Kihansi Spray Toad survival, it was agreed that 25% of the animals released in the Upper Spray Wetland would be marked with a single elastomer tag while none of the animals released in the Mhalala Spray Wetland would receive an elastomer tag.

It was also agreed that in the Upper Spray Wetland half of the animals would be released in the vicinity of Old Frog Rock and the other half in vicinity of New Frog

Fig. 15.1 William Newmark holding an insulated container housing Kihansi Spray Toads on a flight from Dar es Salaam to Kihansi. To his left is Fadhila Khatibu, LKEMP project coordinator, and behind him is Rafik Hirji, World Bank senior water resources specialist

Fig. 15.2 Members of the Tanzanian and World Bank delegation delivering an insulated container of Kihansi Spray Toads from the University of Dar es Salaam captive breeding facility to the Kihansi captive breeding facility

Rock, two locations in the Upper Spray Wetland that prior to and shortly after the commissioning of the dam contained high densities of toads. Toads that were released in the vicinity of Old Frog Rock received a green tag, while those released in the vicinity of New Frog rock received a purple tag.

Between 21 and 29 October, 2500 Kihansi Spray Toads were transferred from the University of Dar es Salaam captive breeding facility to the Kihansi captive breeding facility by road and air (Figs. 15.1 and 15.2).

Early in the morning of 30 October 2012, the Kihansi Spray Toad attendants at the Kihansi captive breeding facility and IUCN consultants began packing 27–28 Kihansi Spray Toads in 32 oz. reptile deli cups. Nine cups were placed in eight Styrofoam chests which in turn were carried up into the gorge by a team of porters (Fig. 15.3). Representatives from NEMC, TAWIRI, TANESCO, Wildlife Division, UDSM, SUA, LKEMP, Bronx and Toledo zoos, local villages, and the World Bank hiked up into the gorge and gathered outside of the Upper Spray Wetland.

The Deputy Director of NEMC, who served as the guest of honor, released the first toads back to the gorge shortly after 2:30 pm. All individuals who hiked up to the Upper Spray Wetland had an opportunity to release one or more reptile deli cups of Kihansi Spray Toads back to the spray wetland, and by 5:00 pm all 2000 Kihansi Spray Toad had been transferred back to the spray wetland.

Everybody who released toads had the same expression—a broad grin (Fig. 15.4). I can remember how elated I felt after releasing the first deli cup of toads. Each deli

Fig. 15.3 Porters transferring the Kihansi Spray Toad from the Kihansi captive breeding facility to the Upper Spray Wetland prior to the first hard release in October 2012

cup of toads represented tens of thousands of hours of combined efforts on the part of multiple organizations and individuals and was truly a cause for celebration.

Shortly after the release of the first Kihansi Spray Toad back to the Kihansi Gorge, Radio Tanzania, the national radio station, interrupted their normal broadcast to announce the reintroduction of the Kihansi Spray Toad back to the Kihansi Gorge. This radio announcement highlighted the pride within Tanzania as well as the high visibility that the Kihansi Spray Toad recovery program had within Tanzania. Over the following week, a series of popular articles appeared in national and international newspapers announcing the reintroduction of the Kihansi Spray Toad back to the Kihansi gorge.

Three days after the release of 2000 Kihansi Spray Toads in the Upper Spray Wetland, Tracey Tuberville, Kurt Buhlmann, Nassoro Mohamed, and Severinus Mutagwaba released 302 unmarked Kihansi Spray Toads in the Mhalala Spray Wetland followed 3 days later by the release of an additional 133 spray toads into the same wetland.

15.2 Post-release Monitoring of the First Hard Release

A post-release monitoring program was jointly developed by the Tanzanian reintroduction committee and the IUCN consultants. It was agreed that a subset of plots that had been established under the IREM project would be monitored initially every 2 weeks over the first 2 months and thereafter monthly. These plots were located in the vicinity of the Old and New Frog Rocks in the Upper Spray Wetland.

Fig. 15.4 Representatives from the University of Dar es Salaam, National Environmental Management Council, Toledo and Bronx zoos, IUCN, and the World Bank participating in the first hard release of the Kihansi Spray Toad in the Upper Spray Wetland. (Photo Credit: Kurt Buhlmann and Tracey Tuberville)

In addition, transect surveys of the Kihansi Spray Toad in the spray wetland vegetation would be conducted every 3 months in the Upper Spray Wetland following the ecological monitoring protocols for amphibians that had been developed for the Kihansi Gorge (Gibbs 2006). Finally, it was agreed that all opportunistic sightings of the Kihansi Spray Toad would also be recorded.

Because of concerns about minimizing the impact of human trampling on the spray wetland vegetation in the Mhalala Spray Wetland—the only spray wetland that had not been impacted by the 90% reduction in flow in the Kihansi River—it was agreed that the Kihansi Spray Toad in this wetland would be less intensively monitored, approximately once per month during the first 2 months and thereafter every 2 to 3 months.

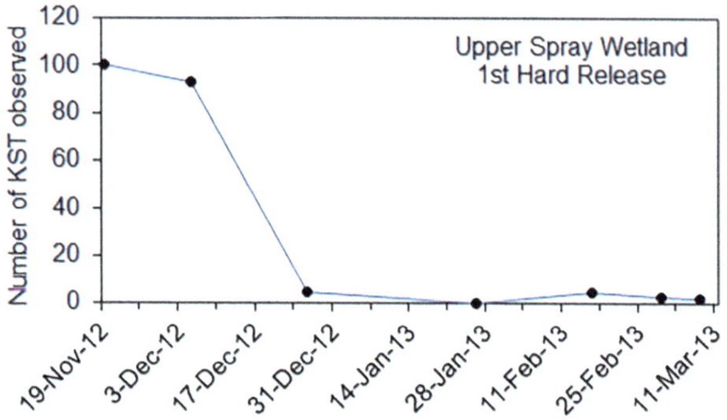

Fig. 15.5 Change over time in the number of Kihansi Spray Toads observed in the Upper Spray Wetland following the first hard release of 2000 toads on 30 October 2012. The number of Kihansi Spray Toads observed per survey includes all animals recorded on rock faces, along vegetation transects, and observed opportunistically. Surveys were conducted during the day and/or night. Number of observers per survey ranged 2–4 individuals, and number of hours of observations per survey ranged 2–4.5 hours. (Source: Msuya 2013; Tuberville and Buhlmann 2014)

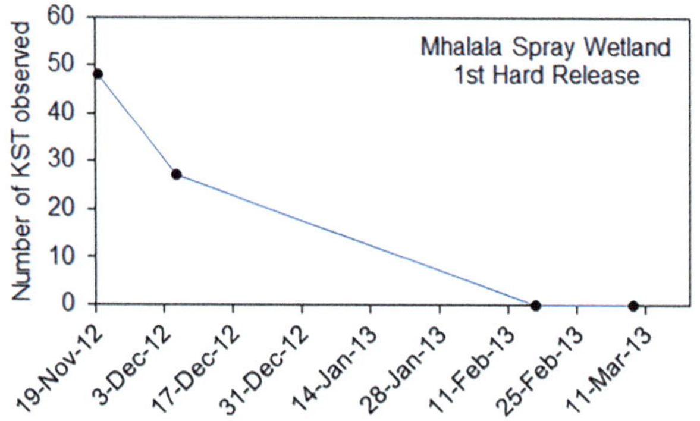

Fig. 15.6 Change over time in the number of Kihansi Spray Toad observed on rock faces and in the vegetation following the first hard release of 435 animals on 2 and 5 November 2012 in the Mhalala Spray Wetland. All surveys were conducted during the day. Number of observers per survey ranged from 2–4, and number of hours of observations per survey ranged from 1 to 1.5 hours. (Source: Msuya 2013; Tuberville and Buhlmann 2014)

Post-release patterns of abundance of the Kihansi Spray Toad were highly similar in the Upper and Mhalala Spray Wetlands. There was a rapid decline in abundance between weeks 2 and 4 followed by a slower decline over the following 3 months (Figs. 15.5 and 15.6).

15.3 Second Hard Release

Tracy Tuberville and Kurt Buhlmann along with Tanzanian counterparts organized a second hard release of the Kihansi Spray Toad in the Upper Spray Wetland between 10 and 14 March 2013. On 10 March 2013, 1050 adults and subadults were released in the Upper Spray Wetland followed by an additional release of 450 adults and subadults on 14 March. Of the 1500 animals that were released, 25% ($n = 375$) were marked with a single elastomer tag.

Representatives from LKEMP, UDSM, SUA, NEMC, TANESCO, local villages, and the World Bank participated in the second hard release of the Kihansi Spray Toad.

15.4 Post-release Monitoring of the Second Hard Release

Based on the knowledge gained from the first hard release in the Upper Spray Wetland of the Kihansi Spray Toad, a more refined post-released monitoring program was developed for the second hard release. Following the second hard release, number of observers and observations hours were more carefully recorded. Thus, changes in abundance of the Kihansi Spray Toad could be expressed in terms of numbers of Kihansi Spray Toads observed per observer-hour. Furthermore, because the Kihansi Spray Toad is more observable at night, all post-release monitoring of the Kihansi Spray Toad following the second hard release was conducted at night at an interval of every 2 weeks over the first 3 months and then thereafter approximately monthly. Skin swabs from live Kihansi Spray Toad as well as dead and live Kihansi Spray Toad specimens were collected twice a month over the first 2 1/2 months. In addition, fecal samples from the Kihansi Spray Toad were collected and inspected at an interval of 2 weeks over a 2-month period following the second hard release.

The change in abundance over time of the Kihansi Spray Toad in the Upper Spray Wetland following the second hard release was similar to that observed following the first hard release. An initial slow decline during the first month weeks was followed by a rapid decline in numbers of observed Kihansi Spray Toad after the second month and then a slower decline in numbers over the following 2 months (Fig. 15.7).

All skin swabs collected from live and Kihansi Spray Toad specimens between March and May 2013 following the second hard release in the Upper Spray Wetland tested negative for chytrid fungus (Makange 2013). In addition, tests conducted of soil samples and leaves during this same time period from the Upper Spray Wetland and from skin swabs from *Afrixalus fornasini* collected from an animal located along the edge of USW also tested negative for chytrid fungus (Makange et al. 2014). Of the fecal samples collected from ten individuals between April and May 2013, three animals contained ova of intestinal worms or intestinal worms.

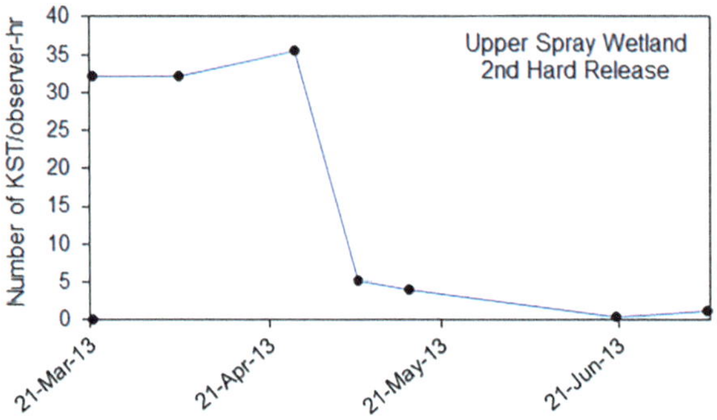

Fig. 15.7 Number of Kihansi Spray Toad per observer-hour following the second hard release of 1500 animals between 10 and 14 March 2013 and 6 July 2013 in the Upper Spray Wetland. All surveys were conducted at night. Number of observers per survey ranged from 2–6, and number of survey hours ranged from 1.5–2. (Source: Mohamed 2013, unpublished data)

Although a rapid decline was observed after the first month during the first two hard releases of the spray toad in the Upper Spray Wetland, an encouraging result, nonetheless, emerged. A small population of the Kihansi Spray Toad following these first two hard releases survived up to 2 years in the Upper Spray Wetland. Between January and November 2014, Nassoro Mohamed and Severinus Mutagwaba recorded during nocturnal surveys in the Upper Spray Wetland, between 2 and 17 individuals during monthly surveys, including gravid females and juveniles. This indicated not only successful reproduction of the Kihansi Spray Toad in the Upper Spray Wetland but also post-reproduction survival of neonates following the first two hard releases in the Upper Spray Wetland.

15.5 Third and Fourth Hard Releases

A third hard release of 1483 Kihansi Spray Toad was conducted in the Upper Spray Wetland on 18 February 2015. Wilirk Ngalason (Fig. 15.8), a herpetologist from the University of Dar es Salaam, organized this release and conducted the post-release monitoring.

The post-release monitoring of the third hard release in the Upper Spray Wetland revealed highly similar declines in numbers of Kihansi Spray Toads over time as had been observed following the first two hard releases. A rapid decline in numbers of Kihansi Spray Toads observed after the first month and followed by a slower decline over the following 3 months. In addition, as had been observed following the first two hard releases, a very small reproducing population survived a minimum of 24 months after the third hard release.

Fig. 15.8 Wilirk Ngalason, a herpetologist at the University of Dar es Salaam, lead the third and fourth hard releases of the Kihansi Spray Toad. He also managed the University of Dar es Salaam captive breeding facility and conducted amphibian surveys in the Kihansi Gorge between 2012 and 2019. (Photo credit: Wilirk Ngalason)

References

Gibbs JP (2006) A detailed ecological monitoring protocol for the Lower Kihansi Gorge ecosystem. Final report to the Lower Kihansi Environmental Management Project

Makange M (2013) Unpublished data. Sokoine University of Agriculture, Morogoro, Tanzania

Makange M, Kulaya N, Biseko E et al (2014) Batrachochytrium dendrobatidis detected in Kihansi spray toads at a captive breeding facility (Kihansi, Tanzania). Dis Aquat Org 111:159–164. https://doi.org/10.3354/dao02775

Msuya C (2013) Monitoring of reintroduced population of Kihansi Spray Toad at Kihansi Gorge wetlands. Kihansi Catchment Conservation and Management Project

Tuberville TD, Buhlmann KA (2014) Final report on Kihansi Spray Toad reintroduction efforts during 2012–2013, including progress report for December 2012–July 2013. Kihansi Catchment Conservation and Management Project

Enclosure Studies

16

Abstract

In October 2016, University of Dar es Salaam consultants constructed three large enclosures, 10 m × 6 m × 2 m, in the Upper and Lower Spray Wetlands. Two of the enclosures were placed in the Upper Spray Wetland, and one enclosure was placed in the Lower Spray Wetland. The purposes of the enclosures were (1) to reduce potential predation on the Kihansi Spray Toad; (2) increase the likelihood of individuals finding mates by reducing the maximum dispersal distance within the release zone; and (3) increase the probability of detecting released animals during post-release surveys. In late 2016, 997 Kihansi Spray Toads were placed in Enclosure 1, and 1074 spray toads were placed in Enclosure 2 in the Upper Spray Wetland. Changes in numbers of Kihansi Spray Toad observed over time in the two enclosures were highly similar to that observed during the first four hard releases with a rapid decline over the first month followed by a slower decline over the following 4–5 months.

In an attempt to understand the potential multiple factors that may be contributing to the decline over time in abundance of the Kihansi Spray Toad following the first four hard releases, Charles Msuya (Fig. 16.1) a consultant to KCCMP and colleagues constructed three large enclosures between October and November 2016 in the Upper and Lower Spray Wetlands. Two of the enclosures were placed in the Upper Spray Wetland, and the third enclosure was erected in the Lower Spray Wetlands (Fig. 16.2).

The enclosures, 10 m × 6 m × 2 m (length × width × height), were constructed with galvanized pipe and covered with a combination of 3 mm and 1.5 mm mesh screen. The enclosures were designed to (1) reduce potential predation on the

W. D. Newmark, *The Kihansi Spray Toad* (Nectophrynoides asperginis): *Crisis Biology in the 21st Century*, Ecological Studies 254, https://doi.org/10.1007/978-3-031-89599-9_16

Fig. 16.1 Charles Msuya, a University of Dar es Salaam consultant to the KCCMP project, lead the enclosure studies in the Kihansi Gorge between 2016 and 2019 and lead the design and construction of the Kihansi captive breeding facility. (Photo credit: Kurt Buhlmann and Tracey Tuberville)

Fig. 16.2 Recently constructed Enclosures 1 and 2 in the Upper Spray Wetland. (Photo credit: Charles Msuya)

Kihansi Spray Toad; (2) increase the likelihood that individuals could find mates by reducing the potential maximum dispersal distance within the release zone; and (3) increase the probability of detecting released animals during post-release surveys. It was also believed that acclimating the Kihansi Spray Toad in enclosures prior to their release in the spray wetlands may enhance spray toad survival in the spray wetlands.

On 28 October 2016, 892 Kihansi Spray Toads from the Toledo zoo were released in Enclosure 1 in the Upper Spray Wetland. This was followed by a release of 1074 Kihansi Spray Toads in Enclosure 2 between 19 November and 10 December 2016 in the Upper Spray Wetland. Of the 1074 Kihansi Spray Toad released in Enclosure 2, 150 Kihansi Spray Toads were from the UDSM captive breeding facility, and 924 Kihansi Spray Toads were from the Bronx zoo. Twenty-five percent of all animals in the two releases were marked with a single elastomer tag.

Changes in numbers of Kihansi Spray Toads observed over time in Enclosures 1 and 2 were highly similar to that observed following the first four hard releases of the Kihansi Spray Toad in the Upper Spray Wetland. In Enclosures 1 and 2, Kihansi Spray Toad abundance declined rapidly after the first month followed by a slower decline in abundance over the following 4–5 months (Figs. 16.3 and 16.4). The last recorded individuals ($n = 6$) in Enclosure 1 were observed 17 months after the initial introduction of 892 Kihansi Spray Toad (Msuya and Mohamed 2018; Ngalason et al. 2018).

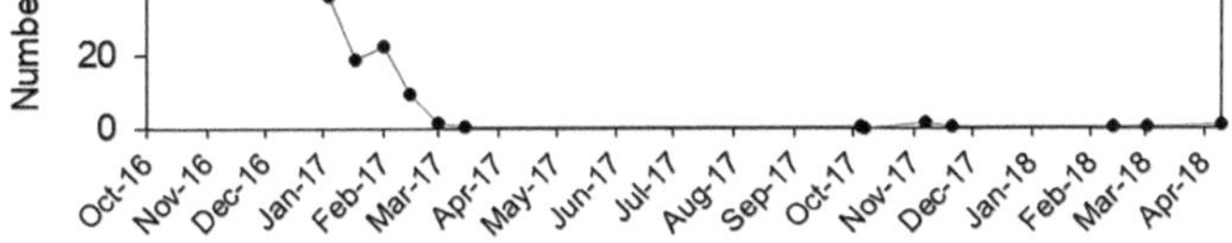

Fig. 16.3 Change in abundance over time of the Kihansi Spray Toad in Enclosure 1 in the Upper Spray Wetland following the release of 892 animals in Enclosure 1 on 28 October 2016. The numbers of Kihansi Spray Toads that were recorded during the first week following release are excluded in this figure because many animals tend to hide in the vegetation immediately following their release. All surveys after the first week of release were nocturnal and were conducted over 1 hour by two individuals walking inside the enclosure along a path that bisected the enclosure lengthwise and then along the outside edge of the enclosure along a path that paralleled the interior path. (Source: Msuya and Mohamed 2017a, b, 2018; Ngalason et al. 2018)

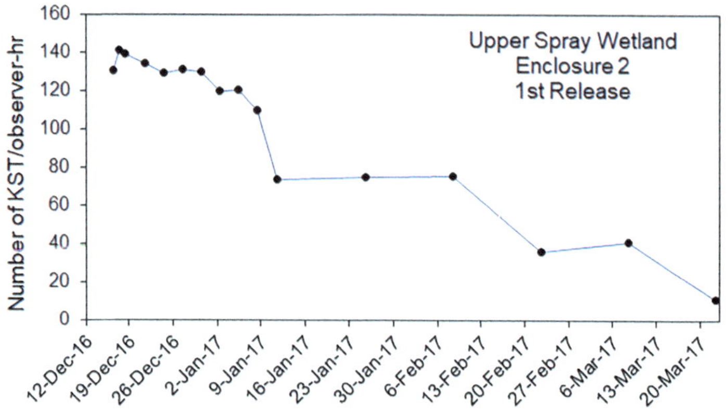

Fig. 16.4 Change in abundance over time of the Kihansi Spray Toad in Enclosure 2 in the Upper Spray Wetland following the release of 1074 animals between 19 November and 10 December 2016. The numbers of Kihansi Spray Toads recorded during the first week following release are excluded in this figure because animals tend to hide in the vegetation immediately following their release. All surveys conducted after the first week of release were nocturnal and were conducted over 1 hour by two individuals walking inside the enclosure along a path that bisected the enclosure lengthwise and then along the outside edge of the enclosure along a path that paralleled the interior path. (Source: Msuya and Mohamed 2017a, b)

References

Msuya C, Mohamed N (2017a) Report on construction and reintroduction of the Kihansi Spray Toad back to its natural habitat by using acclimatizing cages. August–November 2016. Kihansi Catchment Conservation and Management Project

Msuya C, Mohamed N (2017b) Report on reintroduction and monitoring of the Kihansi Spray Toad in its natural habitat and the use of acclimatizing cages. December 2016–March 2017. Kihansi Catchment Conservation and Management Project

Msuya CA, Mohamed N (2018) Progress report on the reintroduction of the Kihansi Spray Toad (KST) to its natural habitat at Kihansi Gorge. Kihansi Catchment Conservation and Management Project

Ngalason W, Nahonyo CL, Ugomba RS et al (2018) Population status of Kihansi Spray Toad (Nectophrynoides asperginis) at the UDSM captive breeding facility. Kihansi Catchment Conservation and Management Project

Mark-Recapture Studies

17

Abstract

A series of mark-recapture studies were conducted between 2017 and 2018 for purposes of assessing multiple hypotheses regarding post-release survival of the Kihansi Spray Toad in the Upper Spray Wetland. In a pilot study conducted in May 2017, 200 of the 590 released animals in Enclosure 2 in the Upper Spray Wetland were individually marked. Animals that were treated for intestinal parasites had higher weekly apparent survival than animals that were not treated although differences were not significant based on an overlap of the 95% confidence intervals. However, unlike in the previous five hard and soft releases in the Upper Spray Wetland, a rapid decline in abundance of the Kihansi Spray was not observed until after month 4. In an expanded study, 701 animals were individually marked and released in May 2018 in Enclosure 2 in the Upper Spray Wetland. In this study animals that were treated for intestinal parasites 6 weeks after release or were treated after both the first and sixth week after release had significantly higher weekly apparent survival than animals that were not treated or were treated only after the first week. This expanded mark-recapture study also revealed no significant difference in weekly apparent survival between (1) animals raised in captivity in the Toledo and Bronx zoos; (2) between gravid and non-gravid females; and (3) by body mass (i.e., age class) after accounting for treatment for intestinal parasites.

In the previous five releases in the Upper Spray Wetland, three hard and two soft, up to 25% of animals had been marked at a cohort level (one color on one leg). However, this non-individual level marking of animals did not permit an estimate of post-release survival nor could competing hypotheses regarding the post-release survival

W. D. Newmark, *The Kihansi Spray Toad* (Nectophrynoides asperginis): *Crisis Biology in the 21st Century*, Ecological Studies 254, https://doi.org/10.1007/978-3-031-89599-9_17

of the Kihansi Spray Toad be evaluated. Additionally, post-release monitoring of fecal samples from animals during the two first soft releases in the Upper Spray Wetland revealed a high intestinal parasite load.

The Tanzanian Kihansi Spray Toad reintroduction committee agreed that individually marking animals and conducting a mark-recapture analysis would be highly beneficial. A pilot study was initiated in May 2017 in Enclosure 2 in the Upper Spray Wetland which involved individually marking 200 of 590 released animals with elastomer tags. The objective of this pilot study was to assess the effect of treatment for intestinal parasites on weekly survival. Animals were monitored weekly over 14 weeks, and all animals both marked and unmarked that were recaptured on weeks 4, 8, and 12 after release were treated for intestinal parasites.

Treatment of intestinal parasites had a positive effect on weekly apparent survival of the Kihansi Spray Toad. Unlike the first five releases in the Upper Spray Wetland, a rapid change in abundance of the Kihansi Spray Toad was not observed until after month 4 (Fig. 17.1).

A mark-recapture analysis was conducted by Monte Neate-Clegg from the University of Utah. Treatment for intestinal parasites in this pilot study was the highest ranked model of weekly apparent survival. Kihansi Spray Toads that were treated 4 weeks after release for intestinal parasites had higher weekly apparent survival than individuals that were not treated. Yet, the difference in weekly apparent survival between animals that were treated in comparison to those that were not was not significant based on an overlap of the 95% confidence intervals (Fig. 17.2).

While the results of the pilot study were suggestive as to potentially important factor(s) influencing post-release survival of the Kihansi Spray Toad in the Upper

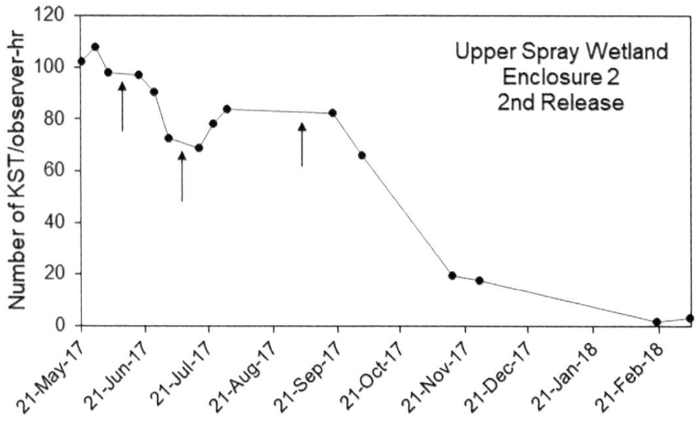

Fig. 17.1 Change in abundance over time of the Kihansi Spray Toad in Enclosure 2 in the Upper Spray Wetland following the second release of 590 animals in Enclosure 2 on 15 May 2017. Two hundred animals were individually marked. Number of observers was 2–4, and observation hours were 1.0–2.0 across all surveys. Three of the 19 surveys are excluded from the figure because a concerted effort was made during these three surveys to capture and treat as many animals as possible for intestinal parasites. Arrows highlight the dates in which animals were captured and treated for intestinal parasites. (Source: Mohamed, *unpublished data*)

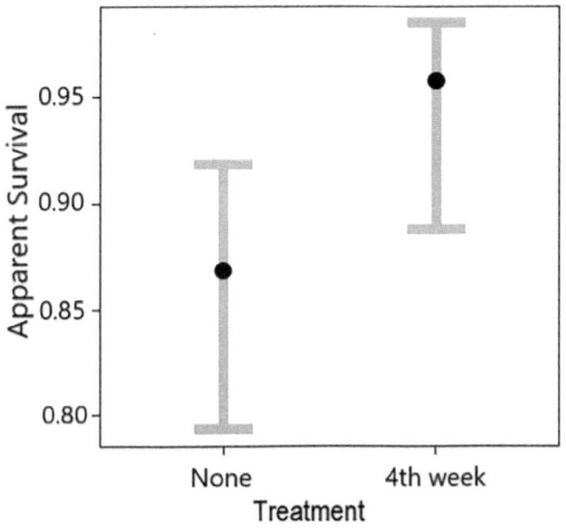

Fig. 17.2 Comparison in a pilot mark-recapture study in Enclosure 2 in the Upper Spray Wetland of mean (±95% C.I.) weekly apparent survival between Kihansi Spray Toads that were treated 4 weeks after release and those that received no treatment. Two hundred of the 590 spray toads that were released in Enclosure 2 in May 2017 were individually marked. Surveys were conducted weekly over the 8-week analysis period

Spray Wetland, they highlighted the importance of individually marking a larger number of animals.

Nassoro Mohamed and two University of Dar es Salaam spray toad attendants, therefore, individually marked 701 adults and subadults. All of the animals were released in Enclosure 2 on 5 May 2018 (Fig. 17.3) and were monitored over 5 months at an interval of 1 to 2 weeks. In this expanded study, a larger number of covariates were examined. These included (1) treatment for intestinal parasites (no treatment; treatment after the first week of release; treatment after the sixth week of release; and treatment after both the first and sixth week after release); (2) age class (adult versus subadult); (3) population source (Toledo versus Bronx zoo); (4) female condition (gravid versus non-gravid); and (5) body mass (i.e., age class).

This multi-model mark-recapture analysis, also conducted by Monte Neate-Clegg, indicated that the highest ranking model of weekly apparent survival among 126 competing models was treatment for intestinal parasites. Animals that were treated for intestinal parasites 6 weeks after release or were treated after both the first and sixth weeks after release had significantly higher weekly survival than animals that received no treatments or animals that were treated only after the first release (Fig. 17.4).

Additionally, the mark-recapture analysis revealed no significant difference in weekly apparent survival between Kihansi Spray Toads (1) raised in captivity in the Toledo and Bronx zoos (Fig. 17.5); (2) gravid and non-gravid females (Fig. 17.6); or (3) by body mass (i.e., juveniles, subadults, adults) after accounting for treatment for intestinal parasites (Fig. 17.7).

Fig. 17.3 Kihansi Spray Toad attendants releasing 701 individually marked Kihansi Spray Toads in Enclosure 2 in the Upper Spray Wetland in May 2018. (Photo credit: Charles Msuya)

Fig. 17.4 Comparison of the effect of treatment for intestinal parasites on weekly apparent survival of the Kihansi Spray Toad. Kihansi Spray Toads that were treated after the sixth week or after the first and sixth week had significantly higher mean (±95% C.I.) weekly apparent survival than spray toads that were not treated or were treated only after the first week

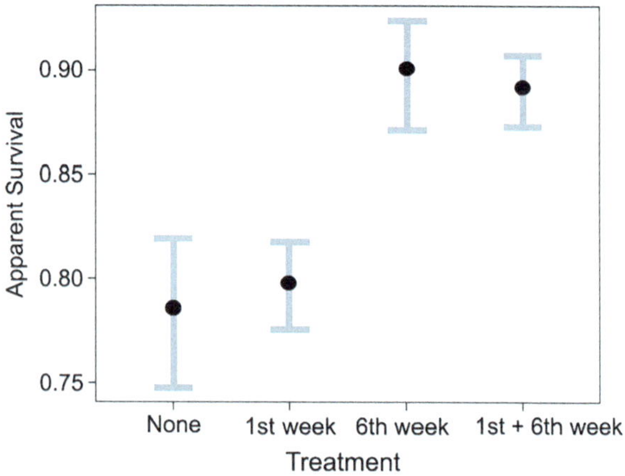

Fig. 17.5 Comparison in Enclosure 2 in the Upper Spray Wetland of mean (±95% CI) weekly apparent survival between Kihansi Spray Toads raised in the Toledo and the Bronx zoos

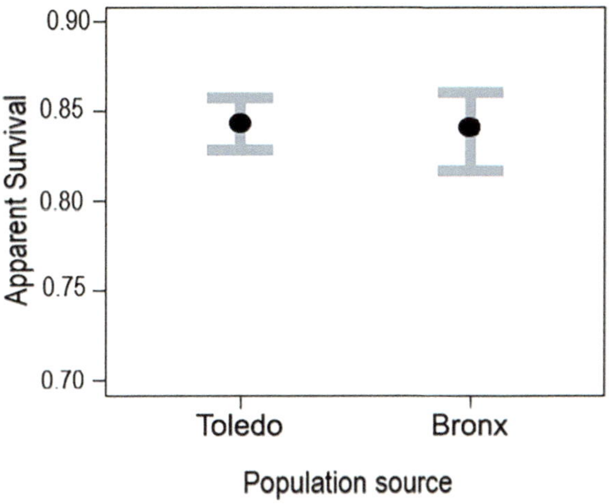

Fig. 17.6 Comparison in Enclosure 2 in the Upper Spray Wetland of mean (±95% CI) weekly apparent survival between non-gravid and gravid females

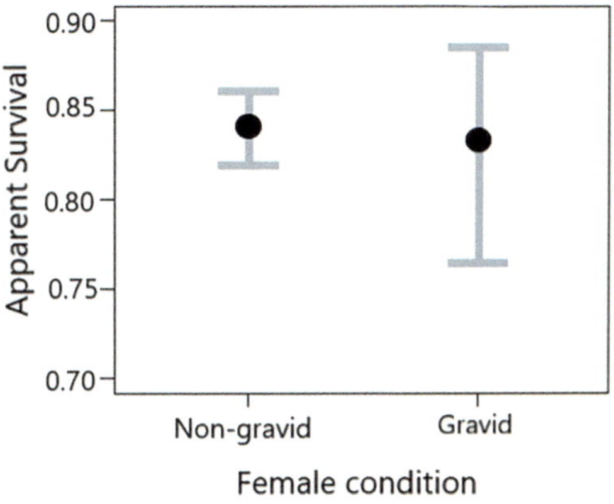

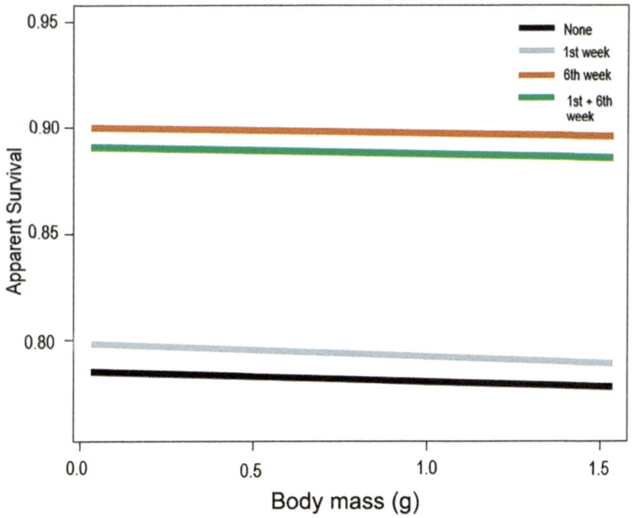

Fig. 17.7 Comparison in Enclosure 2 in the Upper Spray Wetland of the effects of timing of treatment for intestinal parasites on weekly apparent survival of the Kihansi Spray Toad by body mass. Differences in weekly apparent survival by body mass are not significant after accounting for timing of treatment for intestinal parasites as indicated by identical slopes among treatments

Summary of Soft and Hard Releases, 2012–2023

18

Abstract

Between June 2012 and February 2023, 21,474 Kihansi Spray Toads were reintroduced to the Upper, Mhalala, Lower, and Mid-Gorge Spray Wetlands in 42 separate releases. The reintroduction strategy was to release as many animals as possible in multiple reintroduction attempts in order to demographically and genetically rescue the very small reproducing populations of the Kihansi Spray Toad that were surviving up to 2 years in one or more of the spray wetlands. Although it was envisioned during the development of the reintroduction plan for the Kihansi Spray Toad that up to 5000 animals could be reintroduced annually, slightly less than 2000 animals were on average reintroduced annually due to issues related to water quality in the captive breeding facilities and the logistical challenges associated with transferring large numbers of animals between the United States and Tanzania and organizing the releases. The largest number of animals and releases occurred in the Upper Spray Wetland followed by the Mid-Gorge, Lower, and Mhalala Spray Wetlands. The most frequent age class(es) that were released were adults + subadults, followed by babies, adults + babies, adults, subadults, and adults + subadults + babies.

Between June 2012 and February 2023, 21,474 Kihansi Spray Toads were reintroduced to the Upper, Mhalala, Lower, and Mid-Gorge Spray Wetlands in the Kihansi Gorge in 42 separate soft and hard releases (Table 18.1). Over this 11-year period, an effort was made to repeatedly reintroduce large numbers of animals in separate reintroductions to rescue demographically and genetically the very small reproducing populations of the Kihansi Spray Toad that were persisting up to 2 years following many of the soft and hard releases in one or more of the spray wetlands.

W. D. Newmark, *The Kihansi Spray Toad* (Nectophrynoides asperginis): *Crisis Biology in the 21st Century*, Ecological Studies 254, https://doi.org/10.1007/978-3-031-89599-9_18

Table 18.1 Summary of soft and hard releases conducted between October 2012 and February 2023 in the Upper, Mhalala, Lower, and Mid-Gorge Spray Wetlands in the Kihansi Gorge

Spray wetland	Date of release	Number of animals released	Type of release	Enclosure/ no. cages	Age class	Post-release monitoring	Duration of post-release monitoring (months)
Upper	26 Jun 2012	48	Soft	3 × 1 m^3 cages	Adults, subadults, babies	Yes	4
Upper	30 Oct 2012	2000	Hard		Adults + subadults	Yes	24
Upper	6 Nov 2012	48	Soft	3 × 1 m^3 cages	Adults	Yes	4
Upper	10 & 14 Mar 2013	1500	Hard		Adults + subadults	Yes	19
Upper	6 Apr 2013	48	Soft	3 × 1 m^3 cages	Subadults	Yes	4
Upper	6 Apr 2013	10	Soft	1 × 1 m^3 cage	Babies	Yes	4
Upper	18 Feb 2015	1483	Hard		Adults + subadults	Yes	24
Upper	3 Jul 2016	696	Hard		Adults + subadults	No	
Upper	28 Oct 2016	892	Soft	Enclosure 1	Adults + subadults	Yes	5
Upper	19 Nov & 11 Dec 2016	1043	Soft	Enclosure 2	Adults + subadults	Yes	5
Upper	15 May 2017	590	Soft	Enclosure 2	Adults + subadults	Yes	9
Upper	2 May 2018	106	Soft	Enclosure 1, 1 × 1 m^3 cage	Babies	No	
Upper	5 May 2018	701	Soft	Enclosure 2	Adults + subadults	Yes	4.5
Upper	7 May 2018	500	Soft	Enclosure 1	Adults + subadults	Yes	2
Upper	6 Jul 2018	252	Soft	Enclosure 1	Adults + subadults	Yes	2
Upper	23 Jan 2019	300	Soft	Enclosure 1	Adults + subadults	No	
Upper	24 Jan 2019	950	Soft	Enclosure 2	Adults + babies	No	

(continued)

Table 18.1 (continued)

Spray wetland	Date of release	Number of animals released	Type of release	Enclosure/ no. cages	Age class	Post-release monitoring	Duration of post-release monitoring (months)
Upper	14 May 2019	1014	Soft	Enclosure 1	Adults + subadults	No	
Upper	20 Aug 2019	150	Hard		Adults	No	
Mhalala	2 & 5 Nov 2012	435	Hard		Adults + subadults	Yes	4
Lower	26 Jun 2012	48	Soft	3 × 1 m^3 cages	Adults	Yes	4
Lower	6 Nov 2012	48	Soft	3 × 1 m^3 cages	Subadults	Yes	4
Lower	6 Apr 2013	10	Soft	1 × 1 m^3 cage	Babies	Yes	4
Lower	6 Jul 2018	488	Soft	Enclosure 3	Adults + subadults	No	
Lower	27 Oct 2018	50	Soft	Enclosure 3, 1 × 1 m^3 cage	Babies	No	
Lower	27 Oct 2018	50	Hard		Babies	No	
Lower	7 Jan 2019	646	Hard		Adults + subadults	No	
Lower	11 Mar 2020	300	Hard		Adults + subadults	No	
Lower	11 Mar 2020	400	Hard		Babies	No	
Mid-Gorge	27 Oct 2018	150	Hard		Adults + babies	No	
Mid-Gorge	27 Oct 2018	50	Soft	1 × 1 m^3 cage	Babies	No	
Mid-Gorge	15 Jun 2019	400	Hard		Adults + babies	No	
Mid-Gorge	20 Aug 2019	350	Hard		Babies	No	
Mid-Gorge	15 Jan 2020	200	Hard		Adults	No	
Mid-Gorge	15 Jan 2020	300	Hard		Babies	No	
Mid-Gorge	11 Mar 2020	800	Hard		Adults + subadults	No	
Mid-Gorge	11 Mar 2020	218	Hard		Babies	No	

(continued)

Table 18.1 (continued)

Spray wetland	Date of release	Number of animals released	Type of release	Enclosure/ no. cages	Age class	Post-release monitoring	Duration of post-release monitoring (months)
Mid-Gorge	16 Jun 2020	800	Hard		Adults + subadults	No	
Mid-Gorge	16 Jun 2020	400	Hard		Babies	No	
Mid-Gorge	23 May 2021	800	Hard		Adults + subadults	No	
Mid-Gorge	28 Feb 2022	1000	Soft	10 × 1 m^3 cages	Adults, subadults, babies	Yes	5.5
Mid-Gorge	16 Feb 2023	1200	Soft	12 × 1 m^3 cages	Adults, subadults, babies	Yes	6.0
Total		**21,474**					

Source: Tuberville and Buhlmann 2014; Ngalason 2015, 2016; Msuya and Mohamed 2017, 2018; Ngalason et al. 2017, 2018; Msuya 2019; Mtui et al. 2025, N Mohamed personal communication, D Mtui personal communication, P Ugomba personal communication

This reintroduction strategy was adopted because previous meta-analyses that have examined reintroduction success across multiple taxa have found reintroduction success is positively associated with numbers of animals that are released (Griffith et al. 1989; Fischer and Lindenmayer 2000). However, reintroduction success would also appear to asymptote with the numbers of animals released (Griffith et al. 1989; Morris et al. 2021). Among amphibians, projects that have released >1000 animals have had higher success than those that have released <1000 animals (Germano and Bishop 2009). Additionally, multi-taxa meta-analyses have also revealed that multiple reintroduction events can contribute to maintaining high genetic diversity in introduced populations (Uller and Leimu 2011).

In 2010, when the Kihansi Spray Toad reintroduction plan was being developed, it was thought that up to 4000–5000 animals could be released annually. This was due to the large numbers of newborn animals that were being held back from development in the Toledo zoo between 2010 and 2012—up to 5900 animals per year—as a result of concerns about the captive population of the Kihansi Spray Toad exceeding the carrying capacity of the captive breeding facility.

However, the actual number of Kihansi Spray Toads that were released annually to spray wetlands in the Kihansi Gorge between 2012 and 2023 averaged slightly less than 2000 animals per year. Issues related to water quality in the US and Tanzanian captive breeding facilities and zoos as well as the logistical challenges of transferring large numbers of animals between the United States and Tanzania and organizing multiple reintroductions all contributed to fewer number of animals being reintroduced annually than had been originally planned.

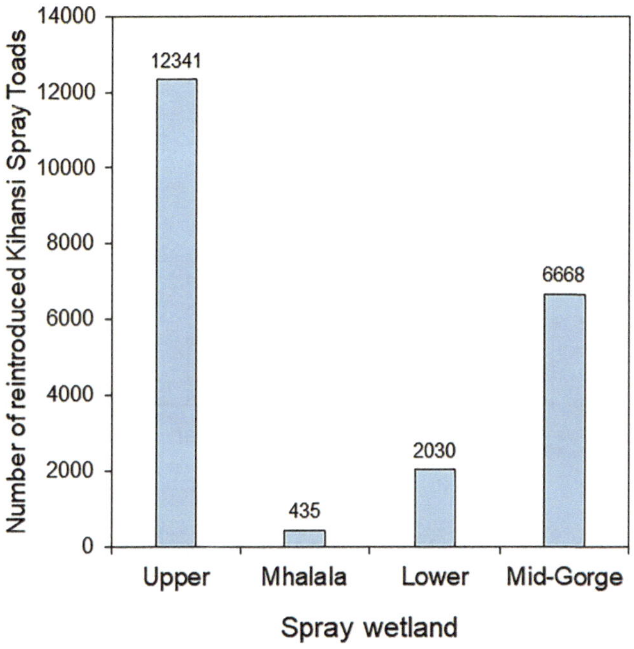

Fig. 18.1 Number of Kihansi Spray Toads reintroduced into the Upper, Mhalala, Lower, and Mid-Gorge Spray Wetlands between June 2012 and February 2023

18.1 Comparison by Spray Wetland

Between 2012 and 2023, the largest number of Kihansi Spray Toads was released in the Upper Spray Wetland followed by the Mid-Gorge, Lower, and Mhalala Spray Wetlands (Fig. 18.1). The largest number of releases was also conducted in the Upper Spray Wetland followed by the Mid-Gorge, Lower, and Mhalala Spray Wetlands, respectively (Fig. 18.2).

18.2 Comparison by Age Class and Post-release Monitoring

Across the 42 individual releases, the most frequently released age classes were a combination of adults + subadults, followed by babies, adults + babies, adults, subadults, and adults + subadults + babies (Fig. 18.3).

Of the 21,474 Kihansi Spray Toads that were reintroduced between June 2012 and February 2023, 55% ($n = 11{,}856$) of all animals were monitored following their release. The mean (±SD) length of time that reintroduced populations was monitored following their release was 7.2 (±6.9) months.

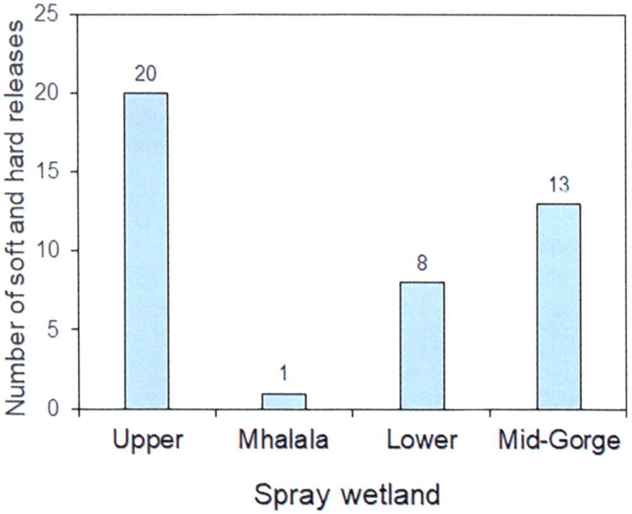

Fig. 18.2 Number of soft and hard releases of the Kihansi Spray Toads conducted in the Upper, Mhalala, Lower, and Mid-Gorge Spray wetlands between June 2012 and February 2023

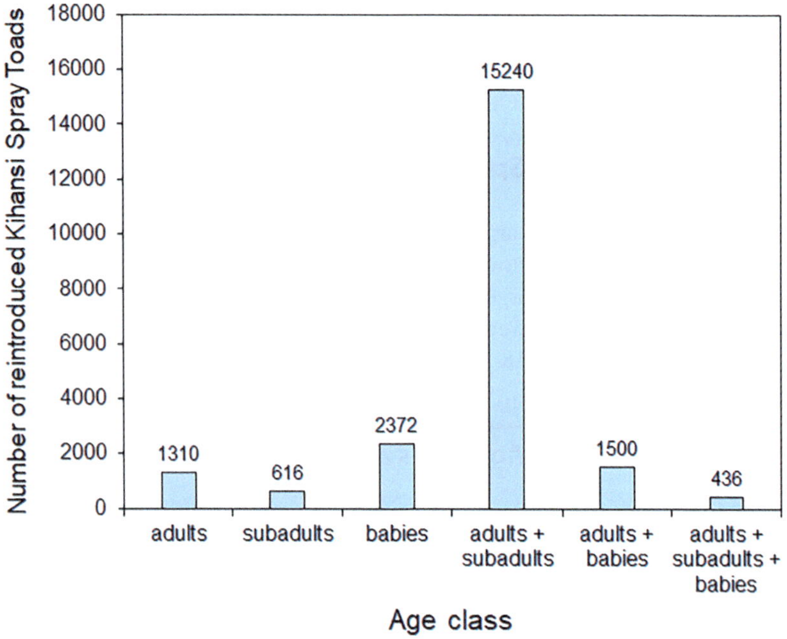

Fig. 18.3 Number of reintroduced Kihansi Spray Toads by age class. In most of the soft and hard releases conducted between June 2012 and February 2023, age classes of animals that were released were combined as adults + subadults, adults + babies, or adults + subadults + babies

A slightly lower percentage of individual releases was monitored. Of the 42 separate releases that were conducted between June 2012 and February 2023 in the Kihansi Gorge, 45% (n = 19) were monitored following the release. Most of the releases that were not monitored occurred between July 2018 and May 2021 as a result of insufficient numbers of personnel and funds to conduct nocturnal surveys toward the end of the KCCMP project.

References

Fischer J, Lindenmayer DB (2000) An assessment of the published results of animal relocations. Biol Conserv 96:1–11. https://doi.org/10.1016/S0006-3207(00)00048-3

Germano JM, Bishop PJ (2009) Suitability of amphibians and reptiles for translocation. Conserv Biol 23:7–15. https://doi.org/10.1111/j.1523-1739.2008.01123.x

Griffith B, Scott JM, Carpenter JW, Reed C (1989) Translocation as a species conservation tool: status and strategy. Science 245:477–480. https://doi.org/10.1126/science.245.4917.477

Morris SD, Brook BW, Moseby KE, Johnson CN (2021) Factors affecting success of conservation translocations of terrestrial vertebrates: a global systematic review. Glob Ecol Conserv 28:e01630. https://doi.org/10.1016/j.gecco.2021.e01630

Msuya CA (2019) Report on the reintroduction of the Kihansi Spray Toad (KST) to its natural habitat at Kihansi Gorge. Kihansi Catchment Conservation and Management Project

Msuya C, Mohamed N (2017) Report on reintroduction and monitoring of the Kihansi Spray Toad in its natural habitat and the use of acclimatizing cages. December 2016—March 2017. Kihansi Catchment Conservation and Management Project

Msuya CA, Mohamed N (2018) Progress report on the reintroduction of the Kihansi Spray Toad (KST) to its natural habitat at Kihansi Gorge. Kihansi Catchment Conservation and Management Project

Mtui D, Haule L, Ogutu J et al (2025) Monitoring the prevalence and distribution of chytrid fungus (Batrachochytrium dendrobatidis) in the Kihansi Spray Toad (Nectophrynoides asperginis) population in the Kihansi Gorge spray wetlands, Tanzania. Authorea in review. https://doi.org/10.22541/au.173720494.40494260/v1

Ngalason W (2015) Proposed monitoring activities/timetable during the third hard release of the Kihansi Spray Toad (Nectophrynoides asperginis) to the Kihansi Gorge. Kihansi Catchment Conservation and Management Project

Ngalason W (2016) Population status of the re-introduced Kihansi Spray Toad Nectophrynoides asperginis. Powerpoint file. Kihansi Catchment Conservation and Management Project

Ngalason W, Nahonyo CL, Msuya CA (2017) The dynamics of re-introduced Kihansi Spray Toad Nectophrynoides asperginis and other amphibians in Kihansi Gorge, Udzungwa Mountains, Tanzania. Powerpoint presentation. Kihansi Catchment Conservation and Management Project

Ngalason W, Goboro E, Nkombe E et al (2018) Assessment of population status of Kihansi Spray Toad Nectophrynoides asperginis and other amphibian species in Kihansi Gorge, Udzungwa Mountains, Tanzania. Kihansi Catchment Conservation and Management Project

Tuberville TD, Buhlmann KA (2014) Final report on Kihansi Spray Toad reintroduction efforts during 2012–2013, including progress report for December 2012–July 2013. Kihansi Catchment Conservation and Management Project

Uller T, Leimu R (2011) Founder events predict changes in genetic diversity during human-mediated range expansions. Glob Chang Biol 17:3478–3485. https://doi.org/10.1111/j.1365-2486.2011.02509.x

Implementation Costs

19

Abstract

The average annual cost between 2002 and 2019 for implementing the species recovery program for the Kihansi Spray Toad was $404,590. The largest single expenditure was the establishment and operation of the captive breeding facilities in the United States and Tanzania, followed by the monitoring of key indicators, chytrid fungus, environmental contaminants, and the reintroduced populations of the Kihansi Spray Toad. The third and fourth largest expenditures were the development of a landscape conservation plan for the Kihansi catchment and the maintenance and expansion of the infrastructure within the Kihansi Gorge. The training of national scientists and technicians and convening of workshops and scientific meetings were the fifth and sixth largest expenditures, followed by the development of a communication strategy for the species recovery program, the seventh largest expenditure.

Few captive breeding-reintroduction programs have reported their associated financial costs (Fischer and Lindenmayer 2000). Yet, the high costs of implementing captive breeding- reintroduction programs have repeatedly been cited in multiple meta-analyses of species reintroductions (Fischer and Lindenmayer 2000; Brady et al. 2017; Berger-Tal et al. 2020) as well as in the IUCN Global Reintroduction Perspective case studies (Soorae 2010, 2011, 2013, 2016, 2018, 2021) as either the largest challenge or one of the major impediments to the implementation of captive breeding-reintroduction programs.

The average annual cost for implementing the Kihansi Spray Toad recovery program between 2002 and 2019 was $404,590. This average annual cost is less than that of several other high-profile species reintroduction and recovery programs such

W. D. Newmark, *The Kihansi Spray Toad* (Nectophrynoides asperginis): *Crisis Biology in the 21st Century*, Ecological Studies 254, https://doi.org/10.1007/978-3-031-89599-9_19

as the California Condor, Sea Otter (Fischer and Lindenmayer 2000), and Bull Trout (Cornwall 2018). On the other hand, the annual expenditure associated with implementing the Kihansi Spray Toad recovery program was higher than that reported for the few amphibian captive breeding-reintroduction programs for which costs have been presented (Brannelly et al. 2013), as well as the estimated hypothetical budget for amphibian reintroductions in Europe and North America (Griffiths et al. 2007).

The single largest expenditure in the Kihansi Spray Toad recovery program was the establishment and operation of the captive breeding facilities and the maintenance of the captive populations at the Toledo and Bronx zoos in the United States and at the University of Dar es Salaam and Kihansi in Tanzania (Fig. 19.1a). The second largest expenditure was ecological monitoring in the Kihansi Gorge which included the costs associated with reintroducing the Kihansi Spray Toad back to the gorge, post-release monitoring of the reintroduced populations, as well as the monitoring of key ecological indicators, chytrid fungus, and other pathogens and contaminants in the Kihansi Gorge. The third largest expenditure between 2002 and 2019 was associated with the development of a landscape conservation plan for the Kihansi catchment to maintain water quality and quantity in the catchment and which included the mapping of all water source points in the catchment. The fourth largest expenditure was maintenance and expansion of the infrastructure in the Kihansi Gorge including the primary and backup spray irrigation systems, research station, bridges, and stairs. The training of national scientists and technicians and convening of workshops and scientific meetings were the fifth and sixth largest expenditures, respectively, followed by the development of a communication strategy for the species recovery program, which was the seventh largest expenditure.

Under LKEMP and KCCMP, many other related activities that contributed indirectly to the conservation of the Kihansi Spray Toad and the implementation of the species recovery program were also supported (Fig. 19.1b). The largest of these related expenditures was the development of an updated Environmental Management Plan, which is a legal requirement for the operation of the Lower Kihansi hydropower facility. The Environmental Management Plan defined the environmental mitigation measures that TANESCO is required to implement as a condition for the operation of the hydropower facility. This expenditure also included the costs of an external audit of the Lower Kihansi hydropower facility to assess whether the environmental mitigation and monitoring activities were achieving their objectives.

The second largest related expenditure conducted under LKEMP and KCCMP was the development and implementation in the Kihansi catchment of an integrative pest management program at the village and district level. This activity was combined at the village level with the support for micro-projects whose objectives were to promote water quality and quantity in the Kihansi catchment and local livelihoods.

The third through eighth largest related expenditures were project management, enhancing capacity in water resource management at the University of Dar Salaam, institutional strengthening of governmental ministries and training in environmental management, conducting a survey of the status of other endangered plant and animal species [*Coffea kihansiensis* (Davis and Mvungi 2004), *Charaxes mtuiae* (Collins et al. 2017)] in the Kihansi gorge, and project communication.

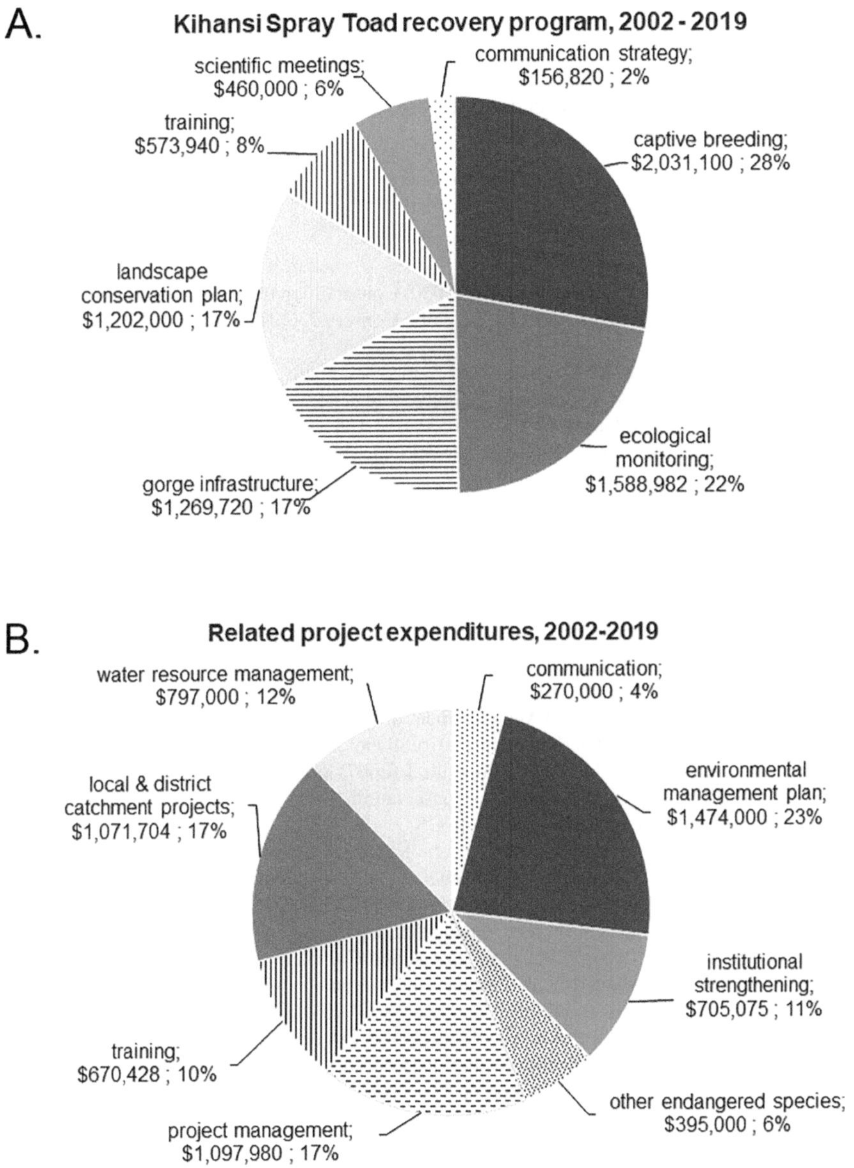

Fig. 19.1 Expenditures between 2002 and 2019 for (**A.**) the Kihansi Spray Toad recovery program and (**B.**) other related project expenditures. Expenditures are presented for those activities having a budgetary line

The Lower Kihansi Environmental Management Project (LKEMP) and the Kihansi Catchment Conservation and Management Project (KCCMP) formally ended the end of June 2011 and December 2019, respectively. External evaluations of LKEMP and KCCMP that were conducted by the World Bank Independent

Evaluation Group were released in April 2014 and May 2020, respectively (Independent Evaluation Group 2014, 2020). Since the formal closing of KCCMP, the government of Tanzania has continued to support the University of Dar es Salaam and Kihansi captive breeding facilities and the reintroduction of the Kihansi Spray Toad back to the Kihansi Gorge.

References

Berger-Tal O, Blumstein DT, Swaisgood RR (2020) Conservation translocations: a review of common difficulties and promising directions. Anim Conserv 23:121–131. https://doi.org/10.1111/acv.12534

Brady L, Young RP, Goetz M, Dawson J (2017) Increasing zoo's conservation potential through understanding barriers to holding globally threatened amphibians. Biodivers Conserv 26:2735–2749. https://doi.org/10.1007/s10531-017-1384-y

Brannelly LA, Chatfield MWH, Richards-Zawacki CL (2013) Method of identification in adult anurans. Herpetol J 23:125–129

Collins S, Congdon C, Bampton I (2017) Review of the Charaxes gallagheri complex resulting in two new species and undescribed morphs of Charaxes gallagheri van Son, 1962 (Lepidoptera, Nymphalidae, Charaxinae). Entomol Africana 22:19–30

Cornwall W (2018) Should it be saved? Science 361:962–965. https://doi.org/10.1126/science.361.6406.962

Davis AP, Mvungi EF (2004) Two new and endangered species of Coffea (Rubiaceae) from the Eastern Arc Mountains (Tanzania) and notes on associated conservation issues. Bot J Linn Soc 146:237–245. https://doi.org/10.1111/j.1095-8339.2004.00328.x

Fischer J, Lindenmayer DB (2000) An assessment of the published results of animal relocations. Biol Conserv 96:1–11. https://doi.org/10.1016/S0006-3207(00)00048-3

Griffiths RA, Buhlmann K, McKay JE, Tuberville T (2007) Reintroductions. In: Gascon C, Collins JP, Moore RD et al (eds) Amphibian Conservation Action Plan: Proceedings: IUCN/SSC Amphibian Conservation Summit 2005. IUCN

Independent Evaluation Group (2014) Lower Kihansi Environmental Management Project Implementation Completion Review. https://documents.worldbank.org/en/publication/documents-reports/documentdetail/517861474594324823/tanzania-lower-kihansi-environmental-management-project. Washington, D.C.

Independent Evaluation Group (2020) Implementation Completion Report – Kihansi Catchment Conservation and Management Project. https://documents.worldbank.org/en/publication/documents-reports/documentdetail/269251594239239925/tanzania-kihansi-catchment-conservation-and-management-project. Washington, D.C.

Soorae PS (ed) (2010) Global re-introduction perspectives: additional case-studies from around the globe. IUCN/SSC Re-introduction Specialist Group, Abu Dabai

Soorae PS (ed) (2011) Global Re-introduction perspectives: 2011. Case studies from around the globe. IUCN

Soorae PS (ed) (2013) Global re-introduction perspectives: 2013. Case studies from around the globe. IUCN

Soorae PS (ed) (2016) Global re-introduction Perspectives: 2016. Case studies from around the globe. IUCN

Soorae PS (ed) (2018) Global re-introduction Perspectives: 2018. Case studies from around the globe. IUCN

Soorae PS (ed) (2021) Global re-introduction Perspectives: 2021. Case studies from around the globe. IUCN

Potential Contributory Factors to Low Post-release Survival of the Kihansi Spray Toad

20

Abstract

Multiple factors may have contributed to the failure to establish a self-sustaining population of the Kihansi Spray Toad in the Kihansi Gorge between 2012 and 2023. The more likely factors include genetic adaptation for captivity, a small founder population, reduced immunological experience, and the continuing presence of *Batrachochytrium dendrobatidis* in the Kihansi Gorge. The effect of these factors could be additive as well as interactive. However, given that the preponderance of evidence suggests that chytridiomycosis was the proximate cause of the extinction in the wild of the Kihansi Spray Toad, highest priority in the future should be given initially to assessing the effectiveness of recently proposed mitigation measures to control chytridiomycosis in situ combined with enhancing the immunological experience of animals prior to reintroduction. This could include exposing animals to low virulent isolates of *Bd* and other native pathogens prior to reintroduction; reintroducing animals to spray wetlands where reservoir host species are absent or their abundance can be controlled; and creating microhabitats in enclosures in the spray wetlands where the Kihansi Spray Toad can potentially self-clear chytridiomycosis.

The continuing presence within and adjacent to enclosures in the Upper Spray Wetland of a very small reproducing population of the Kihansi Spray Toad, $\lesssim$25–50 animals, up to 2 years following their releases between 2013 and 2023 was very encouraging. However, none of these populations were self-sustaining which highlighted the importance of continuously reintroducing animals back to the Kihansi Gorge in an effort to rescue these small populations.

W. D. Newmark, *The Kihansi Spray Toad* (Nectophrynoides asperginis): *Crisis Biology in the 21st Century*, Ecological Studies 254, https://doi.org/10.1007/978-3-031-89599-9_20

Thus the establishment of a long-term self-sustaining population of the Kihansi Spray Toad in the Kihansi Gorge—the overarching objective of the Kihansi Spray Toad recovery program—was not achieved by 31 December 2019, when the KCCMP project ended. Under IUCN's Global Reintroduction Perspective Categories, the reintroduction of the Kihansi Spray Toad back to the Kihansi Gorge between 2012 and 2023 was "partially successful" because of evidence of post-release reproduction and survival of offspring greater than 1 year within and adjacent to enclosures in the Upper Spray Wetland.

Multiple factors may have contributed to the failure to establish a self-sustaining population of the Kihansi Spray Toad in the Kihansi Gorge between 2012 and 2023. Among the more likely contributory factors are genetic adaptation for captivity, a small initial founder population, reduced immunological experience, and the continuing presence of chytrid fungus in the Kihansi Gorge. The effects of these contributory factors on reintroduction success could be additive as well as interactive.

20.1 Genetic Adaptation for Captivity

One of the more challenging issues facing the reintroduction of captive populations to the wild is genetic adaptation to captivity. Genetic adaptation to captivity has been detected in all outbreeding species that have been studied (Frankham 2008). This is a result of alleles that are rare and deleterious in the wild being favored in captivity (Frankham 2008; Robert 2009). Consequently captive-born populations often have reduced survival when reintroduced to the wild in comparison to wild populations (Frankham 2008; Christie et al. 2012; Crates et al. 2023). The extent of adaptation of populations to captivity is dependent in turn on number of generations in captivity, effective population size, genetic diversity, and selection intensity (Frankham 2008).

Minimizing the number generations in captivity is one of the easier means to minimize genetic adaptation to captivity (Frankham 2008; Robert 2009; Williams and Hoffman 2009). However, for many endangered species including the Kihansi Spray Toad, this is often challenging. As of 2023, the Kihansi Spray Toad has been held in captivity for 23 years or approximately 23 generations. In captive steelhead (*Oncorhynchus mykiss*) populations in North America, selection of traits for captivity that were detrimental to survival in the wild was found to occur after one generation (Christie et al. 2012), and among steelhead stocks raised in hatcheries and introduced to the wild, reproductive success declined exponentially by ~40% per generation (Araki et al. 2007). Reduction in genetic heterozygosity and allelic richness as a result of captivity have also been observed in captive amphibian populations. Kraaijeveld-Smit et al. (2006) reported a significant reduction in genetic heterozygosity and allelic richness in captive populations in the Mallorcan midwife toad (*Alytes muletensis*) held for 9 to 12 generations in comparison to those held for 1 to 2 generations.

Reducing genetic adaptation to captivity in threatened species can also be enhanced by housing the captive population across multiple institutions and allowing small effective population size (N_e) and genetic drift to reduce genetic diversity in individual populations while simultaneously maintaining genetic diversity in the entire captive population through the exchange of individuals between institutions (Frankham 2008; Williams and Hoffman 2009). The Kihansi Spray Toad was housed during the first 3 years of captivity in the United States in five zoos. After 2003 all surviving Kihansi Spray Toads were transferred back to two zoos, the Bronx and Toledo, and during the following 6 years, while husbandry protocols were being refined, animals were regularly exchanged between the two zoos. Following the establishment of the University of Dar es Salaam and Kihansi captive breeding facilities in Tanzania, captive-bred animals were moved between these two captive breeding facilities, although most of this exchange was the transfer of animals from the University of Dar es Salaam facility to the Kihansi captive breeding facility. Thus, in theory movement of animals among the US and Tanzanian captive breeding facilities may have helped to reduce the loss of genetic diversity over time in the captive Kihansi Spray Toad population.

Howell et al. (2020) have suggested that supplementing captive-bred populations with biobanked founder sperm may also offer a solution to the challenge of genetic adaptation to captivity and inbreeding depression in captive populations. The biobanking of sperm can also greatly reduce the required colony size and maintenance costs over time. Using the reported captive breeding costs for the Oregon spotted frog (*Rana pretiosa*), Howell et al. (2020) estimate that to retain 90% genetic heterozygosity biobanking combined with backcrossing could reduce costs over 100 years by a factor of 25.6 ($20,992,164 vs $537,142,813) in comparison to managing a captive population without biobanking and backcrossing.

Although integrating assisted reproductive technologies (ART) into captive breeding programs are still in their infancy for amphibians (Clulow et al. 2019; Silla and Byrne 2019; Strand et al. 2020; Calatayud et al. 2022) as well as for most vertebrate species in general (Howard et al. 2016), one of the more high-profile examples where assisted reproductive technologies have been successfully integrated into the captive breeding programs for endangered species is the Black-footed ferret (*Mustela nigripes*). The Black-footed ferret recovery program in the United States was started in 1985 with 18 founders, and sperm was collected and frozen at this time from 6 of the 18 founders (Howard et al. 2016). With recent advances in ART, it was possible after 20 years in captivity to add lost heterozygosity back to the Black-footed ferret population through the cryopreserved sperm. This has resulted in an increase in genetic diversity and lower measures of inbreeding within the captive population, which is now more than 8000 individuals (Holt 2016; Howard et al. 2016).

One of first examples of assisted reproductive technologies beginning integrated into a captive breeding program for amphibians is the endangered Yellow-legged frog (*Rana muscosa*). The San Diego Zoo Wildlife Alliance has recently initiated as part of their long-term ex situ conservation program for the Yellow-legged frog an ART program which includes administration of exogenous hormones, use of artificial fertilization, and sperm cryopreservation (Calatayud et al. 2022). This has resulted in an increase in reproductive output and health of their captive population

(Calatayud et al. 2022). On the other hand, recent experiments with cryopreservation of sperm from Fowler's toad *Anaxyrus fowleri* found that tadpole and post-metamorphic toadlets were smaller than their natural counterparts and that modeling suggested that these early-life differences could scale up to lower fecundity and population trends over time (Poo et al. 2022). Poo et al. conclude that cryopreservation of sperm as a management intervention for enhancing captive populations of threatened amphibian species for translocation is currently problematic and needs additional research.

However, cryopreservation of sperm was not conducted shortly after the Kihansi Spray Toad was brought into captivity. Nor was it possible after the extinction in the wild of the Kihansi Spray Toad in 2006 to enhance the genetic diversity of the captive Kihansi Spray Toad population by transferring additional animals from the wild to captivity. Consequently, reducing selection for captivity in the Kihansi Spray Toad and increasing genetic diversity in the captive populations remains a continuing challenge.

20.2 Small Founder Population

An additional challenge facing the reintroduction of many threatened captive species back to the wild is the loss of genetic diversity and increased inbreeding as a consequence of small founder numbers. Although 499 individuals were transferred in 2000 from the Kihansi Gorge to the Bronx and Detroit zoos, the current captive population of the Kihansi Spray Toad is descended from 71 individuals as a result of the decline of the captive population in US zoos between 2000 and 2004. However, this founding population of 71 is > > 20–30 individuals, the recommended minimum founding population size for captive populations (Frankham et al. 2010). Thus, the current challenges in establishing a self-sustaining population of the Kihansi Spray Toad in the Kihansi Gorge are less likely to be a result of a smaller founder population.

20.3 Reduced Immunological Experience

A further challenge facing many threatened captive species that are reintroduced to the wild is reduced immunological experience due to the control of pathogens while in captivity (Ewen et al. 2012; Stringer and Linklater 2014; Schulte-Hostedde and Mastromonaco 2015; Spencer and Zuk 2016). As a result, reintroduced populations of captive bred species are often vulnerable to formerly benign pathogens.

In order to reduce the likelihood that the captive populations of the Kihansi Spray Toad held in US zoos could transmit exotic pathogens back to Tanzania, the captive populations were housed in biosecure facilities. This housing almost certainly resulted in the immune systems of the captive Kihansi Spray Toad populations being less challenged than that of the historic wild population.

Frequent testing of animals between 2012 and 2018 that were reintroduced back to the spray wetlands revealed high levels of intestinal nematodes. This would suggest that long-term housing of the Kihansi Spray Toad in biosecure facilities in the United States and Tanzania very likely reduced the immunological experience of the Kihansi Spray Toad to formerly benign parasites that occur in the Kihansi Gorge. Indeed, reintroduced spray toads that were treated for intestinal nematodes between 2017 and 2018 had significantly higher weekly apparent survival than untreated animals, although the effects of this treatment on survival were temporary (Fig. 17.1). The long-term housing of the Kihansi Spray Toad in US and Tanzanian captive breeding facilities may have also further weakened the inhibitory response of the Kihansi Spray Toad to chytrid fungus *Batrachochytrium dendrobatidis* (see below). Voyles et al. (2018) reported that the mean inhibitory effectiveness in the variable harlequin frog (*Atelopus varius*) to *Batrachochytrium dendrobatidis* was significantly lower for frogs held in captivity (<10%) than in wild populations (>75%).

The controlled exposure to parasites and other pathogens in a captive environment prior to reintroduction has been proposed as a strategy to build the immune competence of species prior to their reintroduction (Viggers et al. 1993; Faria et al. 2010). Furthermore, pre-exposure in captivity to native parasites can assist in identifying animals that have low resistance to a parasite, and thus if these animals are held back from reintroduction, it can reduce the parasite load in reintroduced populations (Faria et al. 2010).

20.4 Continuing Presence of Chytrid Fungus

A further factor that may have contributed to the repeated decline over time of reintroduced populations of the Kihansi Spray Toad was the continuing presence of *Batrachochytrium dendrobatidis* in the Kihansi Gorge. Although *Bd* was not detected in any of the Kihansi Spray Toads that were tested following the first reintroductions of the spray toad in the Upper and Lower Spray Wetlands between 2012 and 2013, nor was *Bd* detected in any soil or vegetation samples collected during this time period (G. Misinzo *personal communication*, Tuberville and Buhlmann 2014), *Bd* may have nonetheless persisted in putative reservoir host species within the Kihansi Gorge.

In surveys conducted in November 2003, 4 months after the population crash of the Kihansi Spray Toad in the Kihansi Gorge, *Bd* was detected in not only the Kihansi Spray Toad but also in two other anuran species—*Ptychadena anchietae* that were collected adjacent to the Lower Spray Wetland and *Arthroleptides yakusini* that were collected in the Udagaji Gorge, 5 km to the east of the Kihansi Gorge (Weldon and du Preez 2004; Weldon et al. 2020). In a follow-up survey conducted in 2006, *Bd* was again detected in *Arthroleptides yakusini* in animals collected in the Upper Spray Wetland and the Handaki stream which is located above and to the west of the Upper Spray Wetland, as well as in animals collected in the Udagaji Gorge (Weldon and Mtui 2006; Weldon et al. 2020). Ché Weldon concluded that

Arthroleptides yakusini and *Ptychadena anchietae* are very likely reservoir species for *Bd* in the Kihansi Gorge (Weldon and Mtui 2006, Weldon et al. 2020).

More recent surveys conducted in December 2022 for *Bd* in the Kihansi Gorge detected *Bd* in not only Kihansi Spray Toads that had been placed in experimental cages in the Mid-Gorge Spray Wetland but also from skin swabs collected from three other anuran species (*Arthroleptides yakusini*, *Ptychadena anchietae*, *Hyperolius substriatus*) that were sampled within or adjacent to the Upper, Lower, and Mid-Gorge Spray Wetlands (Mtui et al. 2025).

Two of these three putative reservoir species were regularly recorded in amphibian surveys conducted between 2006 and 2022 in one or more of the spray wetlands in the Kihansi Gorge (Table 20.1). The most widespread of the three putative reservoir species, *Arthroleptides yakusini*, was recorded on wet rock faces across all surveys conducted between 2006 and 2022 in one or more of the Kihansi spray wetlands (Mutagwaba 2013; Ngalason et al. 2019). The second most widespread putative reservoir species in the Kihansi Gorge is *Ptychadena anchietae*. This species was recorded in 9 of 12 surveys conducted between 2006 and 2022 in the Kihansi spray wetlands and was most frequently observed in the Lower Spray Wetland (Table 20.1). Given the current distribution of these three putative reservoir species within the Kihansi Gorge, it is thus very likely that *Bd* has persisted since its introduction to the Kihansi Gorge in reservoir host species.

Additionally, it has been documented that *Bd* can persist in locations where amphibian communities have collapsed in non-amphibian hosts such as reptiles, fish, and birds (Fisher and Garner 2020) and can also be transmitted by water, fog, and very likely from plant surfaces (Kolby et al. 2015a, b; Prado et al. 2023). Moreover, reservoir and susceptible host species do not need to be in the same place at the same time for transmission to occur (Burns et al. 2021).

Finally, the likely persistence of *Bd* in the Kihansi Gorge since its introduction has almost certainly been enhanced by the microclimate of the spray wetlands. The mean daily temperature within the Upper, Lower, and Mid-Gorge Spray Wetlands in the Kihansi Gorge (Chap. 13.2) overlaps the optimum temperature for chytrid fungus, 17°–25° C (Fisher et al. 2021).

20.4.1 Proposed Mitigation Measures

Although the success of translocations of amphibians to sites where chytrid fungus is present is very low, a number of mitigation measures have been recently proposed to increase this success (Garner et al. 2016; Scheele et al. 2019, 2021; Brannelly et al. 2021; Knapp et al. 2023; Berger et al. 2024).

One of the more promising approaches is using naturally recovering populations as source populations to repopulate vacant habitats. In a 15-year landscape-scale reintroduction of the Mountain Yellow-legged frog in the Sierra Mountains in California, 9 of 12 populations that were reestablished by introducing animals from source populations that were naturally recovering to vacant habitat successfully reproduced despite the presence of *Bd* in the environment (Knapp et al. 2023). The

Table 20.1 The recorded occurrence in amphibian surveys between 2006 and 2022 of *Arthroleptides yakusini*, *Hyperolius substriatus*, and *Ptychadena anchietae*, three putative reservoir species for *Batrachochytrium dendrobatidis* in the Upper, Mhalala, Lower, and Mid-Gorge Spray Wetlands

Wetland/putative reservoir species	May 2006	Nov 2006	May 2010[a]	Oct 2010[a]	Apr 2013[b]	Feb 2015	May 2015	Dec 2015	Apr 2016	Dec 2017	Feb 2018	Mar 2018	Dec 2022[a]
Upper Spray Wetland													
Arthroleptides yakusini	x		x	x	x	x	x		x	x	x	x	x
Hyperolius substriatus	x							x					x
Ptychadena anchietae		x											x
Mhalala Spray Wetland													
Arthroleptides yakusini						x					x	x	
Hyperolius substriatus													
Ptychadena anchietae													
Lower Spray Wetland													
Arthroleptides yakusini	x	x	x	x			x	x	x	x	x	x	x
Hyperolius substriatus													

(continued)

Table 20.1 (continued)

Wetland/putative reservoir species	May 2006	Nov 2006	May 2010[a]	Oct 2010[a]	Apr 2013[b]	Feb 2015	May 2015	Dec 2015	Apr 2016	Dec 2017	Feb 2018	Mar 2018	Dec 2022[a]
Ptychadena anchietae		x		x		x	x	x		x	x	x	x
Mid-Gorge Spray Wetland													
Arthroleptides yakusini	x	x		x									x
Hyperolius substriatus													
Ptychadena anchietae							x	x	x	x		x	

Source: Weldon and Mtui 2006; Ngalason and Howell 2007; Mtui 2009; Mutagwaba 2010, 2013; Ngalason et al. 2019; Ngalason *personal communication*; Mtui et al. 2025

[a]Survey of amphibians restricted to the Upper, Lower, and Mid-Gorge Spray Wetlands

[b]Survey of amphibians restricted to the Upper Spray Wetland

evidence for reproduction was the presence of tadpoles, juveniles, and new adults at these formerly vacant sites (Knapp et al. 2023). However, this approach is not possible, at least currently, for the Kihansi Spray Toad because of the absence of any naturally recovering populations in the Kihansi Gorge.

An alternative proposed strategy is to immunize captive amphibian populations by exposing animals to low virulent isolates of *Bd* (Greener et al. 2020; Berger et al. 2024). In laboratory experiments midwife toads (*Alytes obstetricans*) exposed to low virulent isolates of *Bd*GLO had significantly higher survivorship when exposed to high virulent isolates of *Bd*GLO than toads that were not pre-exposed to low virulent isolates (Greener et al. 2020). Furthermore, animals that were pre-exposed to low virulent isolates of *Bd*GLO developed less infections when exposed to high virulent isolates of *Bd*GLO than animals that were not pre-exposed to low virulent isolates (Greener et al. 2020). Laboratory experiments with the western North American leopard frog *Rana onca* also found that exposing animals to a hypovirulent strain of *Bd* followed by clearance with an anti-fungal (itraconazole) provided significant resistance to chytridiomycosis when exposed to a virulent isolate of *Bd* over a 12-week period (Waddle et al. 2021). Thus pre-exposing captive populations of the Kihansi Spray Toad to a nonvirulent isolate of *Bd*Cape—the identified strain of chytrid fungus that occurs in the Kihansi Gorge (Sewell et al. 2024)—prior to reintroducing animals back to the Gorge should be explored.

An additional mitigation strategy that has been proposed to increase translocation success of amphibians to sites where chytrid fungus is present is to translocate, remove, or cull reservoir species (McCallum 2012; Garner et al. 2016; Scheele et al. 2021). While there are ethical issues associated with removing or culling native species, the effectiveness of this mitigation strategy has not, as far as I am aware, been evaluated in any field trials. Yet, it may be feasible to test this proposed mitigation strategy in one or more of the smaller spray wetlands in the Kihansi Gorge where numbers and diversity of putative reservoir species are absent or very low.

Bioaugmentation with local *Bd*-inhibiting probiotics has also been proposed as a mitigation strategy to protect *Bd*-naïve amphibian populations as well as increase translocation success of amphibians at sites where chytrid fungus is endemic (Harris et al. 2009; Bletz et al. 2013). Laboratory studies have found that treating the skin of the Mountain yellow-legged frog (*Rana muscosa*) with antifungal bacteria can prevent morbidity and mortality in animals when exposed to *Bd* (Harris et al. 2009). However, in recent field trials in the Sierra Mountains with the Mountain yellow-legged frog, Knapp et al. reported that protective effects from a bioaugmentation treatment of subadult animals declined within 2 months after treatment (Knapp et al. 2022). In this field trial, subadults were initially treated with itraconazole to reduce *Bd* loads and then exposed *Janthinobacterium lividum*, a commensal bacterium with antifungal properties (Knapp et al. 2022). However, in these same field trials, adults that were only treated with itraconazole but not exposed to *J. lividum* did persist over the first 3 years of the study but then declined over the following years. Knapp et al. therefore recommended that the effects of repeated antifungal treatments over several years should be evaluated in future experiments because multiple treatments may possibly be more effective in protecting *Bd*-naïve populations than a single treatment (Knapp et al. 2022). On the other hand, studies with the Mallorcan midwife toad (*Alytes muletensis*) and the Mountain chicken frog (*Leptodactylus fallax*) have found that repeated use of antifungal drugs applied directly to amphibians either prior to or after release have not been effective in providing long-term benefits after the treatment ends (Bosch et al. 2015; Hudson 2016).

Finally, translocating amphibians to sites within their former historic range that are suboptimal for *Bd* has also been proposed as a mitigation strategy to protect *Bd*-naïve amphibians populations (Scheele et al. 2021; Berger et al. 2024; Waddle et al. 2024). A number of amphibian species are known to have self-cleared chytridiomycosis by selecting microhabitats that are unfavorable for *Bd* (Berger et al. 2024; Waddle et al. 2024). Thus, the provision of man-made or natural environmental refugia within amphibian translocation sites where animals can self-clear chytridiomycosis has been suggested as mitigation strategy where *Bd* is present (Berger et al. 2024; Waddle et al. 2024). In the Upper and Lower Spray Wetlands in the Kihansi Gorge, the spatial dispersion of artificial spray could be altered so as to create warmer and drier microhabitats within portions of the enclosures, thus providing sites where the Kihansi Spray Toad could potentially self-clear chytridiomycosis.

20.5 Conclusions Regarding Potential Contributory Factors

It is quite likely that multiple factors have contributed to the low post-release survival of the Kihansi Spray Toad in the Kihansi Gorge. The most likely of these factors are genetic adaptation for captivity, reduced immunological experience, and the continuing presence of chytrid fungus in the Kihansi Gorge. Over the last decade, a number of mitigation measures have been proposed to mitigate these factors. Yet given that the preponderance of evidence to date indicates that the proximate cause of the extinction in the wild of the Kihansi Spray Toad was chytridiomycosis, highest priority should be given initially to assessing the effectiveness of recently proposed mitigation measures to control chytridiomycosis in situ. These include exposing animals to low virulent isolates of *Bd*, reintroducing animals to wetlands in the Kihansi Gorge where reservoir species are absent and/or their abundance can be controlled, and creating microhabitats within the enclosures in the spray wetlands where the Kihansi Spray Toad can potentially self-clear chytridiomycosis. Additionally, given the high intestinal parasitic load that have been observed in reintroduced populations of the spray toad, pre-exposing captive-bred Kihansi Spray Toads to native parasites should also be explored.

It is important that the results of field trials to manage chytridiomycosis in situ in the Kihansi Gorge be shared with the wider scientific community because the challenge of successfully reintroducing captive-bred amphibian species is a global challenge.

References

Araki H, Cooper B, Blouin MS (2007) Genetic effects of captive breeding decline in the wild. Science 318:100–103. https://doi.org/10.1126/science.1145621

Berger L, Skerratt LF, Kosch TA et al (2024) Advances in managing chytridiomycosis for Australian frogs: Gradarius Firmus Victoria. Annu Rev Anim Biosci 12:113–133. https://doi.org/10.1146/annurev-animal-021122-100823

Bletz MC, Loudon AH, Becker MH et al (2013) Mitigating amphibian chytridiomycosis with bioaugmentation: characteristics of effective probiotics and strategies for their selection and use. Ecol Lett 16:807–820. https://doi.org/10.1111/ele.12099

Bosch J, Sanchez-Tomé E, Fernández-Loras A et al (2015) Successful elimination of a lethal wildlife infectious disease in nature. Biol Lett 11:1–4. https://doi.org/10.1098/rsbl.2015.0874

Brannelly LA, McCallum HI, Grogan LF et al (2021) Mechanisms underlying host persistence following amphibian disease emergence determine appropriate management strategies. Ecol Lett 24:130–148. https://doi.org/10.1111/ele.13621

Burns TJ, Scheele BC, Brannelly LA et al (2021) Indirect terrestrial transmission of amphibian chytrid fungus from reservoir to susceptible host species leads to fatal chytridiomycosis. Anim Conserv 24:602–612. https://doi.org/10.1111/acv.12665

Calatayud NE, Jacobs LE, Williams CL et al (2022) Recovering an endangered frog species through integrative reproductive technologies. Theriogenology 191:141–152. https://doi.org/10.1016/j.theriogenology.2022.07.003

Christie MR, Marine ML, French RA, Blouin MS (2012) Genetic adaptation to captivity can occur in a single generation. Proc Natl Acad Sci USA 109:238–242. https://doi.org/10.1073/pnas.1111073109

Clulow J, Upton R, Trudeau VL, Clulow S (2019) Amphibian assisted reproductive technologies: moving from technology to application. In: Comizzoli P et al (eds) Advances in experimental medicine and biology. Springer Nature Switzerland AG, pp 413–463

Crates R, Stojanovic D, Heinsohn R (2023) The phenotypic costs of captivity. Biol Rev 98:434–449. https://doi.org/10.1111/brv.12913

Ewen J, Acevedo-Whitehouse K, Alley MR et al (2012) Empirical consideration of parasites and health in reintroduction. In: Ewen JG, Armstrong D, Parker KA, Seddon PJ (eds) Reintroduction biology: integrating science and management. Wiley-Blackwell, pp 290–335

Faria PJ, van Oosterhout C, Cable J (2010) Optimal release strategies for captive-bred animals in reintroduction programs: experimental infections using the guppy as a model organism. Biol Conserv 143:35–41. https://doi.org/10.1016/j.biocon.2009.06.002

Fisher MC, Garner TWJ (2020) Chytrid fungi and global amphibian declines. Nat Rev Microbiol 18:332–343. https://doi.org/10.1038/s41579-020-0335-x

Fisher MC, Pasmans F, Martel A (2021) Virulence and pathogenicity of chytrid fungi causing amphibian extinctions. Ann Rev Microbiol 75:673–693. https://doi.org/10.1146/annurev-micro-052621-124212

Frankham R (2008) Genetic adaptation to captivity in species conservation programs. Mol Ecol 17:325–333. https://doi.org/10.1111/j.1365-294X.2007.03399.x

Frankham R, Ballou JD, Briscoe DA (2010) Introduction to conservation genetics, 2nd edn. Cambridge University Press

Garner TWJ, Schmidt BR, Martel A et al (2016) Mitigating amphibian chytridiomycoses in nature. Philos Trans R Soc B Biol Sci 371. https://doi.org/10.1098/rstb.2016.0207

Greener MS, Verbrugghe E, Kelly M et al (2020) Presence of low virulence chytrid fungi could protect European amphibians from more deadly strains. Nat Commun 11:1–11. https://doi.org/10.1038/s41467-020-19241-7

Harris RN, Brucker RM, Walke JB et al (2009) Skin microbes on frogs prevent morbidity and mortality caused by a lethal skin fungus. ISME J 3:818–824. https://doi.org/10.1038/ismej.2009.27

Holt WV (2016) The black-footed ferret recovery program: a strong advocate for establishing semen banking programs as support tools for small population welfare. Anim Conserv 19:116–117. https://doi.org/10.1111/acv.12270

Howard JG, Lynch C, Santymire RM et al (2016) Recovery of gene diversity using long-term cryopreserved spermatozoa and artificial insemination in the endangered black-footed ferret. Anim Conserv 19:102–111. https://doi.org/10.1111/acv.12229

Howell LG, Frankham R, Rodger JC et al (2020) Integrating biobanking minimises inbreeding and produces significant cost benefits for a threatened frog captive breeding programme. Conserv Lett 14:1–9. https://doi.org/10.1111/conl.12776

Hudson M (2016) Conservation management of the mountain chicken frog. Ph.D. thesis. University of Kent

Knapp RA, Joseph MB, Smith TC et al (2022) Effectiveness of antifungal treatments during chytridiomycosis epizootics in populations of an endangered frog. PeerJ 9:1–31. https://doi.org/10.7717/peerj.12712

Knapp RA, Wilber MQ, Byrne AQ et al (2023) Evolutionary rescue and reintroduction of resistant frogs allows recovery in the presence of a lethal fungal disease. https://doi.org/10.1101/2023.05.22.541534. 1–14

Kolby JE, Ramirez SD, Berger L et al (2015a) Presence of amphibian chytrid fungus (Batrachochytrium dendrobatidis) in rainwater suggests aerial dispersal is possible. Aerobiologia (Bologna) 31:411–419. https://doi.org/10.1007/s10453-015-9374-6

Kolby JE, Ramirez SD, Berger L et al (2015b) Terrestrial dispersal and potential environmental transmission of the amphibian chytrid fungus (Batrachochytrium dendrobatidis). PLoS One 10:1–13. https://doi.org/10.1371/journal.pone.0125386

Kraaijeveld-Smit FJL, Griffiths RA, Moore RD, Beebee TJC (2006) Captive breeding and the fitness of reintroduced species: a test of the responses to predators in a threatened amphibian. J Appl Ecol 43:360–365. https://doi.org/10.1111/j.1365-2664.2006.01137.x

McCallum H (2012) Disease and the dynamics of extinction. Philos Trans R Soc B Biol Sci 367:2828–2839. https://doi.org/10.1098/rstb.2012.0224

Mtui D (2009) Ecological monitoring program: progress report October 2008 to April 2009. Lower Kihansi Environmental Management Project

Mtui D, Haule L, Ogutu J et al (2025) Monitoring the prevalence and distribution of chytrid fungus (Batrachochytrium dendrobatidis) in the Kihansi Spray Toad (Nectophrynoides asperginis) population in the Kihansi Gorge spray wetlands, Tanzania. Authorea in review https://doi.org/10.22541/au.173720494.40494260/v1

Mutagwaba S (2010) Ecological monitoring program: annual report December 2009–December 2010. Lower Kihansi Environmental Management Project

Mutagwaba S (2013) Summary of post hard release monitoring from March 2013. Kihansi Catchment Conservation and Management Project

Ngalason W, Howell KM (2007) Monitoring of Kihansi Spray Toad population (Nectophrynoides asperginis) in the Kihansi Gorge. Lower Kihansi Environmental Management Project

Ngalason W, Nahonyo CL, Msuya CA (2019) The dynamics of re-introduced Kihansi Spray Toad Nectophrynoides asperginis and other amphibians in Kihansi Gorge, Udzungwa Mountains, Tanzania. Tanzania J Sci 45:584–598

Poo S, Bogisich A, Mack M et al (2022) Post-release comparisons of amphibian growth reveal challenges with sperm cryopreservation as a conservation tool. Conserv Sci Pract 4:1–9. https://doi.org/10.1111/csp2.572

Prado JS, Ernetti JR, Pontes MR, Toledo LF (2023) Chytrid in the clouds: an alternative passive transport of a lethal pathogen for amphibians. Hydrobiologia 2061–2073. https://doi.org/10.1007/s10750-023-05218-2

Robert A (2009) Captive breeding genetics and reintroduction success. Biol Conserv 142:2915–2922. https://doi.org/10.1016/j.biocon.2009.07.016

Scheele BC, Foster CN, Hunter DA et al (2019) Living with the enemy: facilitating amphibian coexistence with disease. Biol Conserv 236:52–59. https://doi.org/10.1016/j.biocon.2019.05.032

Scheele BC, Hollanders M, Hoffmann EP et al (2021) Conservation translocations for amphibian species threatened by chytrid fungus: a review, conceptual framework, and recommendations. Conserv Sci Pract 3(2021):e524

Schulte-Hostedde AI, Mastromonaco GF (2015) Integrating evolution in the management of captive zoo populations. Evol Appl 8:413–422. https://doi.org/10.1111/eva.12258

Sewell TR, Van Dorp L, Ghosh PN et al (2024) Archival mitogenomes identify invasion by the Batrachochytrium dendrobatidis CAPE lineage caused an African amphibian extinction in the wild. Proc Biol Sci 291:20241157

Silla AJ, Byrne PG (2019) The role of reproductive technologies in amphibian conservation breeding programs. Annu Rev Anim Biosci 7:499–519

Spencer HG, Zuk M (2016) For host's sake: the pluses of parasite preservation. Trends Ecol Evol 31:341–343. https://doi.org/10.1016/j.tree.2016.02.021

Strand J, Thomsen H, Jensen JB et al (2020) Biobanking in amphibian and reptilian conservation and management: opportunities and challenges. Conserv Genet Resour 12:709–725. https://doi.org/10.1007/s12686-020-01142-y

Stringer AP, Linklater W (2014) Everything in moderation: principles of parasite control for wildlife conservation. Bioscience 64:932–937. https://doi.org/10.1093/biosci/biu135

Tuberville TD, Buhlmann KA (2014) Final report on Kihansi Spray Toad reintroduction efforts during 2012–2013, including progress report for December 2012–July 2013. Kihansi Catchment Conservation and Management Project

Viggers KL, Lindenmayer DB, Spratt DM (1993) The importance of disease in reintroduction programmes. Wildl Res 20:687–698

Voyles J, Woodhams DC, Saenz V et al (2018) Shifts in disease dynamics in a tropical amphibian assemblage are not due to pathogen attenuation. Science 359:1517–1519. https://doi.org/10.1126/science.aao4806

Waddle AW, Rivera R, Rice H et al (2021) Amphibian resistance to chytridiomycosis increases following low-virulence chytrid fungal infection or drug-mediated clearance. J Appl Ecol 58:2053–2064. https://doi.org/10.1111/1365-2664.13974

Waddle AW, Clulow S, Aquilina A et al (2024) Hotspot shelters stimulate frog resistance to chytridiomycosis. Nature 631. https://doi.org/10.1038/s41586-024-07582-y

Weldon C, du Preez L (2004) Decline of the Kihansi Spray Toad, Nectophrynoides asperginis, from the Udzungwa Mountains, Tanzania. Froglog 62:2–3

Weldon C, Mtui D (2006) Follow-up survey at Kihansi Gorge for the presence of the amphibian chytrid fungus. Lower Kihansi Environmental Management Project

Weldon C, Channing A, Misinzo G, Cunningham AA (2020) Disease driven extinction in the wild of the Kihansi spray toad, Nectophry noides asperginis. African J Herpetol 69:151–164. https://doi.org/10.1080/21564574.2020.1752313

Williams SE, Hoffman EA (2009) Minimizing genetic adaptation in captive breeding programs: a review. Biol Conserv 142:2388–2400. https://doi.org/10.1016/j.biocon.2009.05.034

Lessons Learned

21

Abstract

The captive breeding and reintroduction of globally threatened species are among the most difficult of all applied ecology problems. A number of lessons learned emerged from the captive breeding and reintroduction of the Kihansi Spray Toad of which several are specific to amphibians. These include the challenge of reintroducing a captive-bred amphibian species to a location where chytridiomycosis is present, as well as the potential additive and interactive effects of chytridiomycosis with other important drivers of amphibian declines including habitat loss and fragmentation and toxic compounds. A series other lessons learned emerged that have broad applicability across multiple taxa. These include the importance of (1) a long-term commitment; (2) broad base stakeholder support; (3) enhancing national capacity in biodiverse-rich lesser developed nations; (4) a collaborative, experimentally driven adaptive management approach; (5) a targeted monitoring program; (6) external evaluation; and (7) the inclusion of an exit strategy. Recent meta-analyses of translocation success of captive-bred animals also provide important insights into the design of captive breeding—reintroduction programs for threatened species.

The captive breeding and reintroduction of globally threatened species are among the most challenging of all applied ecological problems. Yet given the dramatic decline of biodiversity worldwide, the captive breeding and translocation of species will inevitably be required to conserve an increasing number of species over the coming decades.

Multiple factors can influence the success of the captive breeding and reintroduction of species. The challenges that were associated with the captive breeding and

W. D. Newmark, *The Kihansi Spray Toad* (Nectophrynoides asperginis): *Crisis Biology in the 21st Century*, Ecological Studies 254, https://doi.org/10.1007/978-3-031-89599-9_21

reintroduction of the Kihansi Spray Toad highlight many of the challenges facing the captive breeding and reintroduction of threatened species in general. Additionally, the captive breeding and reintroduction of the Kihansi Spray Toad also highlight many challenges associated with the captive breeding and translocation of threatened amphibians in particular.

21.1 Class-Specific Lessons

21.1.1 Chytridiomycosis

The reintroduction of amphibians to sites where *Batrachochytrium dendrobatidis* is present is challenging. Amphibians are the most threatened class of vertebrates driven largely by the emergence of the chytridiomycosis panzootic (Scheele et al. 2019). Over the last half century, chytridiomycosis has been responsible for the decline of at least 501 amphibian species including 90 presumed extinctions (Scheele et al. 2019). Indeed, Scheele et al. (2019) describe the amphibian chytridiomycosis panzootic as "the greatest recorded loss of biodiversity attributable to a disease." Since the detection in the late 1990s of chytridiomycosis (Berger et al. 1998) and its etiological agents *Batrachochytrium dendrobatidis* (Longcore et al. 1999) and *B. salamandrivorans* (Martel et al. 2013), chytridiomycosis has been identified in amphibian populations in all continents (Fisher and Garner 2020).

Globally, amphibian species that are large-bodied and range-restricted and occur in the wet tropics have been shown to be particularly vulnerable to chytridiomycosis (Scheele et al. 2019). Two of these three life history traits—range-restricted and occurring in the wet tropics—characterize the Kihansi Spray Toad. Furthermore, chytridiomycosis can interact with and exacerbate the adverse impacts of the other major drivers of amphibian declines including habitat loss and fragmentation, climatic change, invasive species, and toxic compounds (Wake and Vredenburg 2008; Hof et al. 2011; Wake 2012; Campbell Grant et al. 2020).

The Kihansi Spray Toad prior to the population collapse in 2003 had one of the smallest geographic ranges, <0.5 ha, for any tetrapod worldwide. Because of its extremely small geographic range, the Kihansi Spray Toad was highly vulnerable to any catastrophic event and in this case initially environmental perturbation and then subsequently disease. Additionally, the microclimate of the Kihansi Gorge spray wetlands was optimal for *Batrachochytrium dendrobatidis*. The mean daily temperature within the spray wetlands (Sect. 20.4.1) overlaps the optimum temperature for chytrid fungus, 17°–25° C (Fisher et al. 2021).

21.1.2 Translocation Success of Amphibians

To date, the translocation success of amphibians for conservation purposes is considerably lower than that for mammals and birds (Sect. 1.3) due in part to the chytridiomycosis panzootic.

In 2000, establishing a captive population of the Kihansi Spray Toad was controversial. This was due to concerns regarding the ownership of the exported animals as well as the fair and equitable sharing of any benefits that may potentially arise in the future from genetic resources. These were issues that had been highlighted in the International Convention on Biological Diversity and which the government of Tanzania had ratified. Hence a condition for the export of the Kihansi Spray Toad to zoos in the United States was that Tanzania could use the progeny from the captive breeding program in the United States to support future in situ conservation efforts. Following the dramatic collapse in June 2003 of the Kihansi Spray Toad population in the Kihansi Gorge, it became immediately apparent that initiating a reintroduction program for the Kihansi Spray Toad would be required to conserve the spray toad in situ.

The reintroduction of the Kihansi Spray Toad to the Kihansi Gorge was the first reintroduction of an extinct-in-the-wild amphibian species in the tropics. However, the absence of a wild population created additional challenges in terms of enhancing the genetic diversity of the captive populations in the United States and Tanzania as well as sourcing animals for reintroduction from the wild.

21.2 Non-class-Specific Lessons

A series of lessons also emerged from the captive breeding and reintroduction of the Kihansi Spray Toad that have broad applicability to the captive breeding and reintroduction of threatened species in general.

21.2.1 A Long-Term Commitment

Implementing species recovery programs for globally threatened species that involve the captive breeding and reintroduction of animals requires a long-term commitment by all stakeholders. For many threatened species, little is often known about their life history and even less about their husbandry. Following the transfer in December 2000 of 499 Kihansi Spray Toads from the Kihansi Gorge to the Bronx and Detroit zoos, it took 5 years of intensive management to build the captive population of the Kihansi Spray Toad back up to 500 animals following the species' decline in 2002 in the US zoos. This required a careful modification and refinement of husbandry protocols for the Kihansi Spray Toad.

Considerable planning is also required to initiate the reintroduction of highly threatened species. Following the rapid expansion between 2007 and 2009 of the captive Kihansi Spray Toad population in the US zoo to >6400 animals, it took an additional 1 1/2 years before captive breeding facilities in Tanzania were prepared to house the Kihansi Spray Toad and 2 years before the required disease transmission experiments could be conducted. Furthermore, after the development of the reintroduction plan for the Kihansi Spray Toad in early 2010, it took an additional 2 years before any of the pilot studies identified in the reintroduction plan could be initiated.

A long-term commitment is also important in species translocation projects because short- to medium-term population growth and survival rates of translocated species may not be indicative of longer-term demographic rates. For example, in a multi-species reintroduction in Australia involving four species of threatened mammals, all species initially reproduced, and their populations expanded over a 3–16-year period before subsequently declining, in some cases up to 23 years after reintroduction. The long-term declines were thought to be as a result of environmental fluctuations, interspecific competition, and trophic cascades triggered by predator removal (Moseby et al. 2024). Similar patterns of positive short- to medium-term population growth followed by long-term declines have also been reported for two extinct-in-the-wild reptile species on Christmas Island in Australia as a result of a breach in a predator barrier (Emery et al. 2024).

21.2.2 Broad-Based Stakeholder Support

Broad-based stakeholder support is also important in implementing a captive breeding and reintroduction program for threatened species. Under LKEMP and KCMMP, considerable effort was devoted to educating and informing decision-makers within Tanzania, the donor community, and the general public about the species recovery program and the status of the Kihansi Spray Toad and its habitat within the Kihansi Gorge. Guided tours were regularly organized for members of parliament and other government officials of the captive breeding facilities in Tanzania and the United States as well as field site visits to the Kihansi Gorge. In addition, progress reports were provided every 6–12 months to the Tanzanian government and the donor community summarizing the status of project activities as well as the challenges. As part of these progress reports, jointly agreed actions were identified that listed the responsibilities of implementing governmental agencies and non-governmental organizations and a timeline for their implementation.

A communication plan was also developed under LKEMP and KCMMP that was focused on both the general public and school groups and which highlighted the project objectives, activities, and outcomes (LKEMP 2004; Maembe and Khatibu 2009). As part of educating the general public, a targeted training program was developed in integrated pest management and land use practices through farmer field schools in 17 villages in the Kihansi catchment. The objective of this training program was to reduce pesticide use in the Kihansi catchment and to enhance water quality and quantity in the catchment as well as local livelihoods in the Kihansi catchment.

21.2.3 Development of National Capacity

Most species recovery programs that have been developed over the last four decades for threatened species have been conducted in developed countries. However, the majority of globally threatened species occur in biodiversity-rich lesser developed nations. Thus, developing national capacity within biodiversity-rich lesser developed nations to design and manage species recovery programs for threatened species is essential to conserving global biodiversity.

The captive breeding and reintroduction of the Kihansi Spray Toad were the first such effort for any amphibian species in the Afrotropics. Considerable time, effort, and resources were devoted to enhancing national capacity in conservation biology, water resource management, and the design and implementation of a species recovery plan including the establishment and management of amphibian captive breeding facilities and the development and implementation of a reintroduction plan for the Kihansi Spray Toad.

While the captive breeding of globally threatened species is playing an increasingly important role for many zoos particularly in developed countries, the reintroduction of globally threatened species has received comparatively less attention and support. Hence the development of national capacity within biodiversity-rich less developed nations in reintroduction biology will be essential over the next few decades to conserve in situ the increasing numbers of threatened species.

21.2.4 Collaborative, Experimentally Driven, Adaptive Management Approach

Given the multiple threats that were facing the Kihansi Spray Toad in the late 1990s and early 2000s which included habitat loss, a 90% reduction in flow of the Kihansi River through the Kihansi Gorge, a decrease in water quality as a result of changing land use, pesticide use in the Kihansi River catchment, countrywide rolling electrical blackouts in the early 2000s, and the introduction of chytrid fungus into the Kihansi Gorge, incorporating a collaborative, experimentally driven, adaptive management approach in the design and implementation of the species recovery program for the Kihansi Spray Toad was essential.

Between 2000 and 2023, more than 30 national and international organizations participated in designing and implementing the species recovery program for that Kihansi Spray Toad and associated project activities. Institutions and individuals having expertise in key areas of the species recovery program including amphibian biology and husbandry; plant, insect, and aquatic ecology; chytridiomycosis, amphibian disease diagnostics; toxicology detection and assessment; integrated pest management; ecological monitoring; population modeling; disease transmission and detection; reintroduction biology; water resource management and monitoring; irrigation system design and management; and community development played key roles in the design and implementation of the species recovery program.

Because of the multiple and complex threats facing the Kihansi Spray Toad, an adaptive management approach was adopted to guide the implementation of the species recovery program. Adaptive management is a structured decision-making approach for dealing with uncertainty when needing to make recurrent decisions (Walters 1986; Williams et al. 2002; Nichols et al. 2011).

An experimentally driven approach was also incorporated in the design and implementation of the species recovery program for the Kihansi Spray. Important experiments that were conducted included assessing (1) the risk that reintroduced Kihansi Spray Toads posed to other native amphibian species as a result of the Kihansi Spray Toad being initially housed in metropolitan collections while in the United States (Weldon et al. 2011); (2) whether spray wetland vegetation and soil at proposed reintroduction sites in the Kihansi Gorge could transmit chytrid fungus to the Kihansi Spray Toad (Msuya et al. 2012); (3) whether concentrations of endosulfan that were detected in soil, water, and reservoir sediments in the Kihansi reservoir and gorge were toxic to the Kihansi Spray Toad (Gibbs et al. 2011); and (4) whether post-reintroduction survival of the Kihansi Spray Toad varies among age classes and between wetlands (Tuberville and Buhlmann 2014; Mohamed and Magige 2020).

A major challenge in incorporating an experimentally driven approach under a species recovery program for the Kihansi Spray Toad was issues related to small sample sizes, which are often an inherent problem when working with highly threatened species as well as the difficulty in conducting landscape-scale experiments. Additionally, addressing concerns about the impact of marking individual spray toads was also an issue, although 701 spray toads were eventually individually marked which permitted an assessment of multiple previously untested hypotheses regarding the post-release survival of the Kihansi Spray Toad.

21.2.5 Monitoring

Monitoring is central to adaptive management. Considerable effort was devoted under LKEMP to designing a monitoring program that could provide a rapid assessment of short- to medium-term project objectives (Gibbs 2006). Extensive consultations were held with natural resource managers in Tanzania in the selection of indicators to be monitored in the Kihansi Gorge. Additionally, a data management system was developed for the monitoring data, although this data management system was never fully utilized at a level commensurate with its design largely because of problems associated with extracting and formatting raw data from the monitoring reports. Consequently, the ability of the monitoring program to provide "nimble" threat mitigation (Robinson et al. 2020) and a rapid assessment of short-term project objectives was less than originally envisioned. Nonetheless, the monitoring of the top-tier indicators in the Kihansi Gorge played a key role in assessing the effectiveness of mitigation measures that were implemented under the species recovery program for the Kihansi Spray Toad as well as evaluating multiple hypotheses regarding the post-reintroduction survival of the Kihansi Spray Toad in the Kihansi Gorge.

21.2.6 External Evaluations

External and impartial peer-reviewed evaluations are central to science. However, the incorporation of external peer-reviewed evaluations in species recovery programs has been surprisingly only minimally discussed in the rapidly expanding literature focused on captive breeding and reintroduction of species, even though the importance of this activity has been highlighted for nearly a quarter of a century (Kleiman et al. 2000).

External peer-reviewed evaluations of species recovery programs can play an important and valuable role in impartially assessing achievement of objectives as well as recommending future actions. External reviews can provide new perspectives and recommendations and a dispassionate assessment and interpretation of methods and success (Kleiman et al. 2000). Important examples of external and impartial peer reviews of species recovery programs include the American Ornithological Union's Committee for Conservation external review of the species recovery program for the California condor (Walters et al. 2010), the National Research Council's (1992) review of the recovery program for the Hawaiian crow, and the Commonwealth Environmental Research Facility's external review of threatened species recovery planning in Australia (Watson et al. 2011).

Kleiman et al. (2000) recommended that all long-term complex conservation programs including species recovery programs be externally reviewed every 5 years and that internal reviews be conducted annually (Kleiman et al. 2000). During the first year of LKEMP, an external review was conducted of all previous research and monitoring that had been carried out under the LKHP and IREM projects. As part of this external review, recommendations were made for future research and monitoring. These recommendations were extremely valuable in designing the research and monitoring programs that were developed and implemented under the species recovery program for the Kihansi Spray Toad.

During the second year of LKEMP, following the collapse of the Kihansi Spray Toad population in the Kihansi Gorge, the external review panel was again reconvened at the request of the Tanzanian Government to review and recommend future research and monitoring in light of the collapse (Sect. 9.4). The panel's recommendations to establish a cell line and to inform the scientific community about the documented presence of chytrid fungus in the Kihansi Gorge were immediately adopted by the government of Tanzania.

During the first 8 years of LKEMP, two international workshops were convened. The focus of the first workshop was to conduct a population and habitat viability assessment for the Kihansi Spray Toad with the objective of providing input for the development of a species reintroduction plan for the Kihansi Spray Toad (Sect. 11.1). The focus of the second workshop lead by the IUCN/SSC Reintroduction Group and Amphibian Specialist Group was to develop a reintroduction plan for the Kihansi Spray Toad (Sect. 11.3). More than 130 individuals attended the two workshops. Past research and monitoring were again reviewed during these two workshops, and recommendations for future research and monitoring were developed and incorporated into the reintroduction plan for the Kihansi Spray Toad.

Lastly following the completion of LKEMP and KCCMP, external peer-reviewed evaluations were conducted of both projects,which are publicly available (Independent Evaluation Group 2014, 2020).

21.2.7 Exit Strategy

An exit strategy, although uncommon in most species recovery plans (Ruiz-Miranda et al. 2020), was implemented under KCCMP. The purpose of an exit strategy is to define the roles and responsibilities of project counterparts after donor support ends. The exit strategy that was developed under KCMMP defined the ongoing responsibilities for the government agencies and non-governmental organizations responsible for implementing the species recovery program for the Kihansi Spray Toad. As part of this exit strategy, a long-term financial plan was developed to support conservation activities in the Kihansi Gorge and the maintenance of the captive breeding population of the Kihansi Spray Toad in Tanzania (Ngaga 2019). This financial plan, however, has yet to be agreed upon by all stakeholders, and thus long-term financial support for the continuation of conservation activities in the Kihansi Gorge and the maintenance of the captive breeding populations in Tanzania remains unclear.

21.2.8 Predictors of Translocation Success

The difficulty in identifying determinants of translocation success of threatened species as illustrated by the reintroduction of the Kihansi Spray Toad is frequently hindered by small sample sizes and the challenges in conducting large-scale controlled experiments. As a result, ecologists are increasingly conducting meta-analyses for purposes of identifying predictors of translocation success across multiple reintroduction efforts. A summary of previous meta-analyses conducted over the last three decades and which focused on individual classes or a combination of classes of threatened, endangered, and at-risk terrestrial vertebrates is presented in Table 1.1.

Previous meta-analyses of translocation success have found higher translocation success associated with (1) large in comparison to small founder populations, for all classes of terrestrial vertebrates (Griffith et al. 1989; Wolf et al. 1996; Fischer and Lindenmayer 2000; Germano and Bishop 2009); (2) wild-sourced in comparison to captive-bred populations, for predominantly mammals and birds (Griffith et al. 1989; Wolf et al. 1996; Fischer and Lindenmayer 2000; Jule et al. 2008; Rummel et al. 2016; Farquharson et al. 2018); (3) populations translocated to the core of their historic range, for mammals and birds (Griffith et al. 1989; Wolf et al. 1996); (4) populations translocated to sites with good to excellent habitat, for mammals and birds (Griffith et al. 1989; Wolf et al. 1996; White et al. 2012); and (5) soft in comparison to hard releases for mammals, birds, reptiles, and fish (White et al. 2012; Tetzlaff et al. 2019; Resende et al. 2021; Gross et al. 2024).

It should be noted, however, that amphibians have been poorly represented in most meta-analyses that have incorporated multiple taxa. This is due presumably to the fewer numbers of translocations of amphibians in comparison to mammals and birds as well as possible publication biases. Miller et al. (2014) found significant publication bias in the reporting of translocation success for herpetofauna in New Zealand with successful translocations being published considerably more frequently than non-successful translocations.

Furthermore, few meta-analyses have compared translocation success of captive-bred, captive-reared, or wild-sourced populations to wild-resident populations. A notable example, however, is a recent meta-analysis by Gross et al. (2024) which included 821 comparisons from 171 studies representing 9 classes and 101 species and compared one or more measurements of fitness (growth, survival, reproduction) or correlates of fitness (movement, body condition) between translocated and wild-resident populations.

A key finding by Gross et al. was that translocated populations had a 64% decreased odds ratio of success compared to wild-resident animals. This effect was largely independent of taxonomic class or measurement of fitness, although this meta-analysis incorporated only three studies that focused on amphibians. A second important finding was that translocated populations provided with pre-release enrichment prior to translocation performed considerably better than populations that did not receive pre-release enrichment fitness relative to the wild-resident group. Pre-release enrichment was defined as fostering pre-release animals with experienced conspecifics; simulating naturalistic environments in ex situ enclosures; anti-predator training; reducing the length in captivity; and soft releases in which supplementary feeding and/or shelter is provided.

Gross et al., however, caution that as a result of the small number of amphibian translocation studies that have compared the success of translocated populations to wild-control populations, their results may be less applicable to amphibians. They recommend that in the future amphibian translocations incorporate wild-control groups for the donor and/or recipient population and that they be simultaneously monitored.

Additional research is clearly required that is focused on enhancing translocation success of threatened species. And of particular importance is the need for research that is focused on addressing the unique challenges associated with the captive breeding and translocation of threatened amphibian species, the most threatened class of vertebrates.

References

Berger L, Speare R, Daszak P et al (1998) Chytridiomycosis causes amphibian mortality associated with population declines in the rain forests of Australia and Central America. PNAS 95:9031–9036

Campbell Grant EH, Miller DAW, Muths E (2020) A synthesis of evidence of drivers of amphibian declines. Herpetologica 76:101–107. https://doi.org/10.1655/0018-0831-76.2.101

Emery JP, Hollanders M, Valentine L et al (2024) Evaluation of trial reintroductions of two extinct in the wild reptile species on Christmas Island. Anim Conserv 1–14. https://doi.org/10.1111/acv.12940
Farquharson KA, Hogg CJ, Grueber CE (2018) A meta-analysis of birth-origin effects on reproduction in diverse captive environments. Nat Commun 9:1055. https://doi.org/10.1038/s41467-018-03500-9
Fischer J, Lindenmayer DB (2000) An assessment of the published results of animal relocations. Biol Conserv 96:1–11. https://doi.org/10.1016/S0006-3207(00)00048-3
Fisher MC, Garner TWJ (2020) Chytrid fungi and global amphibian declines. Nat Rev Microbiol 18:332–343. https://doi.org/10.1038/s41579-020-0335-x
Fisher MC, Pasmans F, Martel A (2021) Virulence and pathogenicity of chytrid fungi causing amphibian extinctions. Ann Rev Microbiol 75:673–693. https://doi.org/10.1146/annurev-micro-052621-124212
Germano JM, Bishop PJ (2009) Suitability of amphibians and reptiles for translocation. Conserv Biol 23:7–15. https://doi.org/10.1111/j.1523-1739.2008.01123.x
Gibbs JP (2006) A detailed ecological monitoring protocol for the Lower Kihansi Gorge ecosystem. Final report to the Lower Kihansi Environmental Management Project
Gibbs JP, Reeve B, Radell C (2011) Evaluation of toxicity of Endosulfan for Kihansi Spray Toads: direct exposure effects and interactions with nutritional status and infection by pathogenic chytrid fungus. An Addendum to the Final Report to the Lower Kihansi Environmental Management Project
Griffith B, Scott JM, Carpenter JW, Reed C (1989) Translocation as a species conservation tool: status and strategy. Science 245:477–480. https://doi.org/10.1126/science.245.4917.477
Gross IP, Wilson AE, Wolak ME (2024) The fitness consequences of wildlife conservation translocations: a meta-analysis. Biol Rev 99:348–371. https://doi.org/10.1111/brv.13025
Hof C, Araújo MB, Jetz W, Rahbek C (2011) Additive threats from pathogens, climate and land-use change for global amphibian diversity. Nature 480:516–519. https://doi.org/10.1038/nature10650
Independent Evaluation Group (2014) Lower Kihansi environmental management project implementation completion review. https://documents.worldbank.org/en/publication/documents-reports/documentdetail/517861474594324823/tanzania-lower-kihansi-environmental-management-project. Washington, D.C.
Independent Evaluation Group (2020) Implementation completion report – Kihansi catchment conservation and management project. https://documents.worldbank.org/en/publication/documents-reports/documentdetail/269251594239239925/tanzania-kihansi-catchment-conservation-and-management-project. Washington, D.C.
Jule KR, Leaver LA, Lea SEG (2008) The effects of captive experience on reintroduction survival in carnivores: a review and analysis. Biol Conserv 141:355–363. https://doi.org/10.1016/j.biocon.2007.11.007
Kleiman DG, Reading RP, Miller BJ et al (2000) Improving the evaluation of conservation programs. Conserv Biol 14:356–365. https://doi.org/10.1046/j.1523-1739.2000.98553.x
LKEMP (2004) Communication strategy and implementation programme. Lower Kihansi Environmental Management Project
Longcore JE, Pessier AP, Nichols DK (1999) Batrachochytrium dendrobatidis gen. et sp. nov., a chytrid pathogenic to amphibians. Mycologia 91:219–227
Maembe A, Khatibu FA (2009) LKEMP communication strategy (June 2009–December 2010). National Environment Management Council. Lower Kihansi Environmental Management Project
Martel A, Spitzen-Van Der Sluijs A, Blooi M et al (2013) Batrachochytrium salamandrivorans sp. nov. causes lethal chytridiomycosis in amphibians. Proc Natl Acad Sci USA 110:15325–15329. https://doi.org/10.1073/pnas.1307356110
Miller KA, Bell TP, Germano JM (2014) Understanding publication bias in reintroduction biology by assessing translocations of New Zealand's herpetofauna. Conserv Biol 28:1045–1056. https://doi.org/10.1111/cobi.12254

Mohamed N, Magige FJ (2020) Survivorship of the reintroduced Kihansi Spray Toads Nectophrynoides asperginis in a soft release experiment at the Kihansi Gorge, Tanzania. J Educ Humanit Sci 9:111–122

Moseby KE, Read JL, Tuft K, Van der Weyde LK (2024) Influence of interactive effects on long-term population trajectories in multispecies reintroductions. Conserv Biol 38:1–15. https://doi.org/10.1111/cobi.14209

Msuya CA, Misinzo G, Malago J (2012) Ex-situ exposure of Kihansi Spray Toad to Kihansi Gorge spray wetlands substrate. Lower Kihansi Environmental Project

Ngaga YM (2019) Economic financing plan for ensuring the long-term conservation and management of the Kihansi Catchment. Kihansi Catchment Conservation and Management Project

Nichols JD, Koneff MD, Heglund PJ et al (2011) Climate change, uncertainty, and natural resource management. J Wildl Manag 75:6–18. https://doi.org/10.1002/jwmg.33

Resende PS, Viana-Junior AB, Young RJ, Azevedo CS (2021) What is better for animal conservation translocation programmes: soft- or hard-release? A phylogenetic meta-analytical approach. J Appl Ecol 58:1122–1132. https://doi.org/10.1111/1365-2664.13873

Robinson NM, Dexter N, Brewster R et al (2020) Be nimble with threat mitigation: lessons learned from the reintroduction of an endangered species. Restor Ecol 28:29–38. https://doi.org/10.1111/rec.13028

Ruiz-Miranda CR, Vilchis LI, Swaisgood RR (2020) Exit strategies for wildlife conservation: why they are rare and why every institution needs one. Front Ecol Environ 18:203–210. https://doi.org/10.1002/fee.2163

Rummel L, Martínez-Abraín A, Mayol J et al (2016) Use of wild-caught individuals as a key factor for success in vertebrate translocations. Anim Biodivers Conserv 39:207–219. https://doi.org/10.32800/abc.2016.39.0207

Scheele BC, Pasmans F, Skerratt LF et al (2019) Amphibian fungal panzootic causes catastrophic and ongoing loss of biodiversity. Science 1463:1459–1463. https://doi.org/10.1126/science.aav0379

Tetzlaff SJ, Sperry JH, DeGregorio BA (2019) Effects of antipredator training, environmental enrichment, and soft release on wildlife translocations: a review and meta-analysis. Biol Conserv 236:324–331. https://doi.org/10.1016/j.biocon.2019.05.054

Tuberville TD, Buhlmann KA (2014) Final report on Kihansi Spray Toad reintroduction efforts during 2012–2013, including progress report for December 2012–July 2013. Kihansi Catchment Conservation and Management Project

Wake DB (2012) Facing extinction in real time. Science 335:1052–1053. https://doi.org/10.1126/science.1218364

Wake DB, Vredenburg VT (2008) Are we in the midst of the sixth mass extinction? A view from the world of amphibians. Proc Natl Acad Sci USA 105:11466–11473. https://doi.org/10.1073/pnas.0801921105

Walters CJ (1986) Adaptive management of renewable natural resources. MacMillian, New York

Walters JR, Derrickson SR, Fry M, Haig D, Marzluff SM, Wunderle JM, Joseph M (2010) Status of the California Condor (Gymnogyps californianus) and efforts to achieve its recovery. The Auk 127:969–1001. https://doi.org/10.1525/auk.2010.127.4.969

Watson JEM, Bottrill M, Walsh J et al (2011) Evaluating recovery planning for threatened species in Australia. Prepared on behalf of the Department of the Environment, Water, Heritage and the Arts by the Spatial Ecology Laboratory. University of Queensland, Brisbane

Weldon C, Msuya C, Misinzo G, Malago J (2011) Ex-situ disease transmission experiment. Lower Kihansi Environmental Project

White TH, Collar NJ, Moorhouse RJ et al (2012) Psittacine reintroductions: common denominators of success. Biol Conserv 148:106–115. https://doi.org/10.1016/j.biocon.2012.01.044

Williams BK, Nichols J, Conroy MJ (2002) Analysis and management of animal populations. Academic Press, San Diego

Wolf CM, Griffith B, Reed C, Temple SA (1996) Avian and mammalian translocations: update and reanalysis of 1987 survey data. Conserv Biol 10:1142–1154. https://doi.org/10.1046/j.1523-1739.1996.10041142.x

GPSR Compliance

The European Union's (EU) General Product Safety Regulation (GPSR) is a set of rules that requires consumer products to be safe and our obligations to ensure this.

If you have any concerns about our products, you can contact us on ProductSafety@springernature.com

In case Publisher is established outside the EU, the EU authorized representative is:

Springer Nature Customer Service Center GmbH
Europaplatz 3
69115 Heidelberg, Germany

Batch number: 10732044

Printed by Printforce, the Netherlands